Space-Earth Matters

Dr. Surendra Parashar

Dear Karen,

With my complements, for your reading pleasure. Thank you so much for teaching piano to me and others.

Best Regards,

Surendra

Sept. 2021

Suite 300 - 990 Fort St
Victoria, BC, V8V 3K2
Canada

www.friesenpress.com

First Edition — 2021

ISBN
978-1-5255-9584-4 (Hardcover)
978-1-5255-9583-7 (Paperback)
978-1-5255-9585-1 (eBook)

1. SCIENCE, SPACE SCIENCE

Distributed to the trade by The Ingram Book Company

About the Author

SURENDRA PARASHAR learned space technology and applications as he earned a PhD in electrical engineering at the University of Kansas, USA in 1974. Much later in 1998 he obtained an Executive MBA from Queen's University, Canada while being a full-time space worker. He retired from the Canadian Space Agency in 2016 after 31 years of public service. Dr. Parashar is a recipient of Queen's Golden Jubilee Medal in 2002.

For my son Pavan
with love and admiration.

SPACE-EARTH MATTERS

Preface

Human perception of Earth in space is ever changing. Once upon a time there was this belief that our Earth is flat. And then there was the Earth-centric view held for centuries that had our universe move around the Earth. It was only in the 16th century that the actual heliocentric astronomical model was accepted; first presented as a mathematical model by Nicolaus Copernicus, a Polish astronomer, and subsequently supported by empirical observations.

In the Copernicus heliocentric astronomical map, the Earth orbits around our Sun along with the other planets in the solar system. This scientific perspective was reinforced in the public imagination through the Apollo space program when in 1968 there was a photo taken from a lunar orbit of Earthrise showing partial Earth just above our moon's surface. And then again in 1972, when there was a photo of Earth's hemisphere called "The Blue Marble" taken from a distance of 29, 000 km from the Earth.

Further in 1990, we all saw the iconic image, the one of a "Pale Blue Dot", our Earth, only half-pixel wide, floating in space, as seen from Voyager 1 spacecraft, from 6.4 billion km out in space. More recently in 2012 two images ("Blue Marble" and "Black Marble") were released, one a detailed digital composite of the western hemisphere and the other a nighttime view of Americas.

Since then Earth continues to be routinely observed and measured with a variety of space-based sensors. With the advent of the Space Age, we can now really look at our Earth from space and appropriately call it Space-Earth; the only known harbour of life in space, a fitting human perception of Earth moving in space through the Cosmos.

Near Earth space offers a unique vantage point to look down and observe the whole Earth in much detail, and unhindered from its atmosphere, to look up to explore our universe. Moreover, travelling through space is necessary for us to explore nearby celestial bodies such as the moon and other planets. As such space now is a beacon for human enterprise through which Earth-orbiting satellites have already become an essential asset in our daily lives.

An inspiring story of the last sixty years is the application of science and technology to further our understanding of Space-Earth and its relationship to other celestial bodies as a part of our universe. This evolution in space exploration and looking back at Earth from space has come about largely due to the sustained push from government agencies with the support from academia through the involvement of industry. The backbone of investments in funding and effort for space has been and remains the public sector.

There are more than eighty countries that have a space program and fourteen countries have the capability to launch an object into space. In recent years increasing private sector investments are being applied to rapidly exploit what space has to offer for commercial purposes. While this is a welcome development, it is worth recalling that space assets offer unique solutions to challenges here on Earth, such as Climate Change, population growth, depletion of natural resources, sustainable food production, preservation of biodiversity, and control of diseases among others.

It is important that we all know about the exploitation of space for public good and are aware that we are all part of space and space is part of us. Just like Earth, space belongs to each of us and it is a common, shared asset for humanity. It is up to civil society globally to ensure that space is employed for the common good, that access to space is available to all and that benefits realized from assets in space are for safeguarding and bettering not only humanity but all life on Earth.

This book is meant as an introduction to Space-Earth, with the main objective to outline how space assets are being applied for sustainable use of natural resources, for safeguarding the health of our home planet and improving the lives on it, and for enhancing knowledge and understanding of our universe. It is hoped that readers will find this endeavour of some benefit in raising their awareness of what space is and what it offers.

This manuscript tries to address the multi-disciplinary scope of space know-how from its evolutionary, scientific history to the basics of space missions, including their orbital mechanics and technological underpinnings. It summarizes the why and how of Earth-orbiting satellites and importantly their applications and utility. And as well it outlines socio-economic benefits resulting from the development, launch and operations of space assets and the framework for their governance.

It gives me great pleasure to thank book editor: Catherine Rupke, cover designer: Colin Parks, and publishing specialist: Mara Owusu. I am particularly grateful to them for making this book attractive and accessible.

Surendra Parashar
November, 2020
South Surrey, B.C. Canada

OUTLINE

1. WHAT IS SPACE?

- Building Blocks of Universe (Matter and Energy), Celestial Bodies and their Motions, Cosmology (Our Universe), Attributes of Space

2. RUDIMENTARY SPACE HISTORY

- Big Bang Theory and Observations, Formation of Stars and Galaxies, Formation of Solar System (Earth and Life forms), Human Enterprise for Ushering Space Age

3. EARTH IN SPACE

- Features of Sun and Planets, Features of Earth (Chemical Elements, Living Organisms, Humanity's March)

4. NEAR EARTH SPACE ENVIRONMENT

- Earth's Atmosphere, Outer Space

5. TRAJECTORIES IN SPACE

- Laws of Gravity and Motion in Space, Planetary Orbital Mechanics (Kepler's Laws), Satellite Orbits around Earth, Satellite Orbital Mechanics

6. SPACE MISSIONS

- Mission Components (Space and Ground Segments), Mission Cycle (Development Phases, Satellite Launch and Operations, Satellite Disposal)

7. TECHNOLOGY FOR SPACE

- Qualifications for Space, Satellite Bus (structure, thermal control, power generation, propulsion, attitude control, navigation, command and data handling), Satellite Payload (sensor resolutions, passive sensors, active sensors), Ground Infrastructure, Robotic Explorations, Human Explorations

8. SPACE APPLICATIONS AND USES

- Telecommunications, Navigation, Earth Observations (Geophysical Information Extraction, Weather, Mapping, Environment Monitoring, Disaster Management), Space Science (Solar Research, Space Environment, Astronomy and Cosmic Research), Space Explorations, Earth-Orbiting Space Laboratories

9. SPACE ECONOMY AND BENEFIT

- Economy of Space (National Expenditures, Commerce, Educations), Global Space Benefits (Space in Daily Use, Societal Benefits)

10. GLOBAL GOVERNANCE

- National Organizations, International Organizations (United Nations)

Annex A: Units, Nomenclature, Images and Videos
Annex B: Canada in Space

1. WHAT IS SPACE?

You and I exist in space. All beings, human and animal, each exists physically in space. We have brains through which we try to make a sense of the world around us, each brain also exists in space. Whatever we sense, everything we see, hear, or touch, each exists in space.

The air we all breath and rely upon to keep us alive and to sustain us, exists in space. Through inhaling and exhaling we share a common experience of air moving through space. Water, the crucial requisite for life, exists in space. Plant and vegetation that give us sustenance, each exists in space.

When we move, our bodies move through space. We all are in space and space is in each of us. We are all built from the same basic material as the rest of our universe in space, namely-matter and energy.

BUILDING BLOCKS OF UNIVERSE - Matter and Energy

We and every other body in space, each is an amalgamation of matter. Matter is what we call any entity that occupies space and we associate a spatial volume or size and a mass with it. The mass to volume ratio is called density; bodies with bigger mass but having the same volume are denser.

According to the natural science of physics, matter is composed of atoms, each atom made up of subatomic particles called protons, neutrons and electrons. Each subatomic particle contains many fundamental or elementary particles grouped as quarks and leptons. Protons and neutrons make up the nucleus of an atom around which electrons move in orbits. An atom's nucleus is bounded or held together through what we call the strong nuclear force. And electrons are bonded to the nucleus through the weaker electromagnetic force.

Elementary and subatomic particles are attributed with size and mass and a nucleus carries most of an atom's size and mass. Protons and neutrons have approximately the same mass, about 1.67×10^{-24} g (-trillionth trillionth of a gram) while electrons only have about 9.11×10^{-28} g; all virtually massless. The radius of a proton is about 1.1×10^{-15} m (millionth billionth of a metre) while the radius of an electron is 2.8×10^{-15} m; each inconceivably minuscule. In comparison, radii of atoms range from 30 to 300×10^{-12} m (trillionth of a metre).

These particles also have another physical attribute we call electric charge with an electric field around it. The electric charges can be negative (electron), positive (proton) or neutral (neutron). Like charges repel and unlike charges attract through their electric fields. Each electron orbit around the nucleus in an atom is associated with an energy level, meaning there is a change in its energy when an electron moves from one orbit to another. Moving electrons result in microscopic electric currents that in turn produce magnetic fields.

When an electron moves to a lower energy orbit, i.e., closer to the nucleus, a packet of electromagnetic radiation or wave energy, called a photon (also viewed as an elementary particle with no mass) is released, and when it is the other way around—an electron moves to a higher orbit away from the nucleus—a photon is absorbed. This concept is part of physics known as quantum mechanics that incorporates quantization of energy and the wave-particle duality of light and deals with interactions between energy (radiation) and matter.

These emitted photons of electromagnetic radiation propagate through space as waves, and are visible light when they are of the frequency or wavelength that our eyes can see. Visible light waves form a band of wavelengths from 700 to 400 nm (nm-nanometer is a billionth of a meter), a wavelength spectrum which corresponds to the order of colours in a rainbow: red, orange, yellow, green, blue, indigo, violet. Shorter wavelengths have greater frequency and higher energy. These light waves or photons illuminate the world around us, and allow us to see matter.

Electromagnetic radiation can be viewed as synchronized oscillations of electric and magnetic fields forming a transverse wave, where the two fields are perpendicular to each other and also perpendicular to the direction of wave propagation. The frequency (number of cycles per second-expressed as unit Hertz-Hz) and wavelength (distance travelled in one cycle) of such a radiation are interlinked through the speed of light; a constant equal to about 300,000 km per sec. The wavelength as such can be determined by dividing the light speed by the frequency.

Light that we see during daytime are photons emitted from our Sun as a part of its radiation that propagates through the intervening space between the Sun and us. It takes about 8.3 minutes for light originating from the Sun to reach Earth. This solar radiation is called radiant (electromagnetic) energy and here on Earth it provides not only illumination and heat but also the energy for plant growth through the process known as photosynthesis.

Light that we see during the night are photons emitted from other stars, like our Sun, that move through intervening space between these stars and us. Like light waves, sound and music that we hear are also vibrations or waves, they are vibrations of air moving through space. Our sky is blue most of the time because blue light from the Sun is scattered in all directions by atmospheric gases and particles. Blue light is scattered more than other colours because blue travels as shorter, smaller waves.

The most basic atom in our universe is that of hydrogen; it has one proton, no neutron and one electron. Hydrogen is the first chemical element of matter; it is the lightest and the most prevalent in the universe. Hydrogen atoms fuse together to make the next element helium and this process of fusing continues until the heavier elements of iron and beyond get created.

Atoms of elements combine to form chemical molecules and biomolecules and further these molecules combine to create biological cells, the smallest units of life, the building blocks of all living organisms. The size of a molecule is around 10^{-9} m (one billionth of a metre) while cells normally range between 1-100 μm (μm-micrometre is millionth of a metre).

Astronomer Carl Sagan is reported to have famously exclaimed that if you want to make an apple pie from scratch, you must first create the universe. Indeed, and he could have easily added that the prerequisite is that space must be created first to host the universe of matter and energy that we experience. Without space, there is no energy or matter, the building blocks of our universe. Without space, there are no atoms to make an apple and no energy to bake a pie, and importantly there is neither you nor I to bake one.

Thus, in our universe, energy like solar radiation is simply a wave or vibration propagating through space. As well, mass of matter is itself viewed as compressed energy; a small mass can release tremendous energy such as in nuclear fusion. The elementary particles that make up matter are unbelievably small and light—almost virtual and massless—exhibiting a fluffiness as if there is nothing there but empty space, in fact 99% of an atom is empty space.

CELESTIAL BODIES AND THEIR MOTIONS

Our home planet, Earth, the only known harbour of life, exists in space. The Earth and its surrounding atmosphere of air revolves around itself in space. We see our moon as it reflects light from our Sun; it moves through space around Earth. Earth and the moon together move through space around our Sun. We not only see the Sun but feel its presence through the heat energy generated from the Sun's photon energy impacting upon Earth's atmosphere. There are seven other planets, and their moons, like the Earth-moon, all move through space around our Sun.

Our Sun itself moves through space around the Milky Way galaxy, our home galaxy, that contains billions of stars like our Sun. Our galaxy is itself moving through space along with billions of galaxies that we see in the night sky. And so, when we gaze up towards our sky we are peering through space and see all the bodies that exist in space and make up our universe.

When we peer into space, we are peering back into time. We are seeing photons that were emitted in the past but are just reaching us, travelling the distance from stars and galaxies in far reaches of our universe. Our Sun as well as other stars also emit electromagnetic radiation of different wavelengths or frequencies than that of visible light, such as longer wavelength infrared or shorter wavelength ultraviolet, that our eyes do not see.

The emission from stars travelling through space also includes high energy electromagnetic radiation consisting of charged particles (called cosmic radiation or rays), the shortest possible wavelength gamma rays, and x-rays with 0.01-10 nm wavelength. Long exposure to these can be harmful to humans and animals. However, we commonly employ without harm longer wave radiations such as microwaves (1mm to 1m wavelength) or radio waves (wavelength >1 m) in terrestrial communications as they are of lower frequency and lower energy.

From subatomic particles to photons to celestial bodies like planets, stars and galaxies, everything in space is in ceaseless motion, a constant change and evolution. In the mid-17th century, an English mathematician and physicist named Sir Isaac Newton ushered in a new scientific era of classic mechanics when he formulated laws that govern motion of particles or bodies in space and of the forces they are subjected to.

Newton's first law of motion, the law of inertia, states a body continues in a state of rest or motion unless acted by an outside force. The second law states that force is equal to mass of body times acceleration (change in velocity, i.e., its speed and direction). Newton's third law states to every action there is an equal and opposite reaction.

Equally important, Newton also formulated the law of gravity that states there is a force of attraction between any two bodies that is proportional to the product of two masses and inversely to the square of the distance between them. When we combine Newton's law of gravity with his second law of motion, we can deduce the acceleration that a body experiences from the force of Earth's gravity (the rate of fall towards Earth), this being independent of body's mass but decreases as the distance from Earth increases.

These laws describe the motion of moons, planets, and other celestial bodies and are applied in space faring by us, humans. In this Newtonian view, space is static, it is simply a stage or a boundless container or medium through which bodies move while their motion is scripted as per Newton's laws.

COSMOLOGY - Our Universe

It was Albert Einstein, a German physicist, who in the early 19th century through his theories of relativity (special and general) radically transformed the notions of space and time and of mass and energy. In his general theory formulations, space is dynamic, not static as viewed by Newton; space and time are not independent but need to be considered together as interwoven space-time; that space-time is the fabric or a single continuum of our universe.

Any object with a large mass, like Earth, causes a stretching of space-time fabric, trampoline-like, meaning a well or depression is created around the object. Thus, space-time itself gets curved around a massive object, and this curvature is what gets experienced as gravity's force of attraction. Accordingly, objects near a large body move in a curved space towards that large body.

In Einstein's Special theory formulations, laws of physics are universally the same, and the speed of light is the same for all observers, independent of their relative motions. This means that two events occurring at the same time to one observer may not appear simultaneous to another observer. The speed of light (C) in a vacuum is a constant and nothing in our universe can travel faster than this speed. Moreover, energy (E) and mass (M) are equivalent and transmutable through the famous $E= MC^2$, and mass-energy together are conserved in any transaction.

Cosmology or the nature of our universe as postulated by Einstein's theoretical formulations is increasingly validated through observations and measurements. Light from distant stars is seen to bend or deflect around our Sun due to its massive gravity. Regions in space are discovered, called black holes, where nothing escapes their gravity's pull, even light. When these dense objects with their massive gravities merge or combine, gravity waves which are essentially ripples in space-time, are inevitably produced. These gravity waves have recently been detected and measured.

Our universe is seen to be expanding; it is enlarging at an accelerated pace. This expansion is attributed to what is called dark energy—dark, as it does not interact with the matter and energy we can detect, nor is there enough visible mass to hold a galaxy together through gravity's pull. Thus, it is postulated that there is dark matter that keeps a galaxy together.

It is estimated that about 96% of our universe is made up of dark energy and dark matter. While our universe is considered to be around 13.8 billion years old, the expanse of our observable universe is vast, estimated to be about 93 billion lightyears in diameter, i.e., a distance that a photon of light transverses in a time period of 93 billion years.

ATTRIBUTES OF SPACE

English word *space* is derived from the Latin word *spatium*, meaning expanse, and later the French word *espace*. The word *space* seems to have been first used by John Milton, an English poet, in 17th century in his epic poem "Paradise Lost" to indicate just some place in describing the Cosmos. Outer Space is the name now generally applied to define space beyond 100 km above Earth. At this altitude, Earth's atmosphere is sparse and thus this altitude gets viewed as a nominal limit of aircraft, beyond which now fly spacecraft.

So, what is space? We can identify the following attributes:

- Space manifests itself only through the matter and energy in it, just like energy manifests itself only through its interaction with matter.
- Space contains matter, but it also permeates matter, meaning there is a single continuum in space from inside of any subatomic or elementary particle to the farthest reaches of our known universe, an expanse estimated to be around 93 billion lightyears wide.
- Space hosts four known universal forces: strong nuclear that binds the nucleus of an atom, weak nuclear that binds the entirety of an atom, electromagnetic that results in interaction between electrically charged particles, and gravity that pulls matter together.
- Space jiggles as it is a medium for the propagation of electromagnetic and gravity waves, and for the projection of electric, magnetic, and gravity fields.
- Space is spatially three dimensional, whereby three measurements: length, width and height are required to describe an object in it. However, time is considered as the fourth dimension, given that a moment in time is required to place any object in our universe of space-time. Peering through space is equivalent to peering through time, as it shows what the universe was like when the photons that left an object that we now see.
- Space is dynamic, it expands and bends, reacting to matter in it.
- Space is estimated to accommodate more than 100 billion galaxies. Our own Milky Way galaxy hosts around 300 billion stars like our Sun. There are perhaps ten million upwards to a billion black holes in our galaxy alone.

- Space is considered to be essentially flat across our known universe, even though there are curved spaces that result from massive objects, but these are sparse in the vast expanse of space, like pebbles in an ocean of space.
- Space is contributory to the creation of atoms of different elements, to the combination of these elements into molecules and cells and to the coalescing of planet Earth with its biodiversity.
- Space is conducive to production of all ingredients required for the creation of life and the evolution of a variety of life forms.
- Space is supportive to production of a brain, an organ that is self-aware and aware of its surroundings and continues to search for its origins and is curious to understand the why of this creation and the meaning of the Cosmos.

While we continue to learn more and more of our universe in space and study and explore outer space near our Earth, the nature of space itself remains as mysterious and elusive as ever.

2. RUDIMENTARY SPACE HISTORY

Science is the accumulation of human knowledge over time through observations, by the building of hypotheses, theories and models and the testing of their validity through empirical, objective measurements. Scientific knowledge tells us about us, about the world around us, about our universe and our place in it.

This search for the nature of life and the universe springs from our brains' curiosity and creativity and is a shared human quest for non-subjective truth. Scientific methodology is meant to be self-correcting, repeatable, cumulative, transparent and is evidence based. As new evidence arrives, older perspectives are over-hauled or completely replaced.

Science has many disciplines, each devoted to a particular or specialized knowledge of our universe. For example, cosmology deals with the origin and development of the universe while astronomy deals with celestial objects and the physical universe as a whole. Physics is the study of matter and energy, and chemistry deals with the properties of matter and their interactions to make molecules and compounds, while biology is the study of living organisms.

More and more scientific research and technology developments are multi-disciplinary and are multi-faceted endeavours especially when directed towards space.

BIG BANG THEORY AND OBSERVATIONS

Science traces back and tells us that our universe began with what we call a Big Bang that is estimated to have occurred around 13.8 billion years ago. The Big Bang theory is the prevailing accepted model for our observable universe. The Big Bang event is viewed as a singularity, i.e., it sprung forth by itself from a point source. The point source, thought to be smaller than an atom, was extremely dense and hot.

In the very first instant of the Big Bang, space got created along with elementary particles (e.g., quarks and leptons, as well as photons) and the four basic forces (strong and weak nuclear, electromagnetic and gravity) to govern their interactions. This creation of our universe occurred in an infinitesimal small amount of time, estimated to be less than 10^{-37} seconds, and through a period of rapid expansion of space called inflationary.

During the inflationary period the volume of space is estimated to have increased by a factor of at least 10^{78} times from being initially just 10^{-35} m wide. At this time, space was thought to be filled with a uniform glow of white-hot plasma; a universe of photons and particles at temperatures of more than a billion Kelvin (K is the base unit of temperature and can be converted to degrees Celsius).

Ever since the Big Bang, space has been expanding in three dimensions with all the matter and energy in it, and it is cooling as well. The continuing cooling has allowed for conditions to evolve along the way that resulted in the structure of the universe of stars, galaxies, planets and other celestial bodies that we see in space today, all formed from the elementary particles of matter at the start of the Bing Bang.

The initial cooling and then rapid expansion resulted in the formation of first subatomic particles and then atoms. It is estimated that just 10^{-6} seconds after the start of time, quarks (an elementary particle) combined with the help of gluons (another elementary particle) to form neutrons and protons, the nuclei of hydrogen with one proton. Subsequently, a few minutes later, neutrons and protons combined to form nuclei of deuterium (heavy hydrogen—a stable isotope of hydrogen with one proton and one neutron) and helium (two protons with one or two neutrons).

After about 380,000 years the universe was cool enough—around 3,000 K—for electrons and nuclei to combine to form atoms, mostly of hydrogen. This allowed the release of photons that until now were scattered around through collisions with other particles. At this juncture, photons were not just scattering off neutral atoms but began to travel freely through space and these photons have been cooling ever since.

There is a relic called the Cosmic Background Radiation in our universe to mark this epoch. This electromagnetic radiation is known as the Cosmic Microwave Background radiation because it is at a microwave frequency of 160.4 GHz (Giga is billion) or a wavelength of 1.9 mm, and it fills the entire universe.

This important discovery was made in 1965 and subsequently starting in 1989 an entire map of the Cosmic Background Radiation from satellite measurements has been produced. This radiation gives space a residual blackbody temperature of around 2.72 K, just above absolute zero. The measurement of this radiation constitutes one evidence of the validity of the Big Bang theory.

Of the four fundamental forces unleashed in the beginning, it was the three strong forces (strong and weak nuclear and electromagnetic) that were the most influential in ushering in the change until the creation of first atoms. The strong nuclear force is the strongest of the four as it bound together positive charged protons inside of a nucleus. Likewise, the weak nuclear force, only one millionth of the stronger one, is meant to keep a nucleus stable. But it is operative only at a really short distance, about the diameter of a proton.

The electromagnetic force is only 0.7 % as strong as the strong nuclear; while it keeps electrons tied to nucleus within an atom, it has infinite range. Similarly, the force of gravity is the weakest of the four, at about $6x10^{-29}$ times that of the strong nuclear, but has an infinite range. While gravity is only an attractive force, the other three can be attractive or repulsive, each operating through a carrier particle. In our daily lives, we get to only experience the gravity force, from the Earth's pull, and the electromagnetic force from the photons of light.

FORMATION OF STARS AND GALAXIES

With the creation of hydrogen atoms, it was possible for the gravity force to exert its influence. The atoms from a cloud of predominantly hydrogen atoms were able to coalesce together through the gravitational pull into first stars when the universe was thought to be around 200 million years old. Subsequently over time, again through gravity, stars got bound together in to galaxies (collections of billions of stars and gas held together in a unit through gravity) and galaxies into clusters and super clusters. There are thought to be as much as 100 billion galaxies in our observable universe.

There is a diversity of stars, and stars usually get classified into seven main types based on the age or stage of brightness and colour (spectrum of chemical elements that they contain) and temperature. However, they fall into three very different types of stars. Most stars are Main Sequence Stars and are fuelled by nuclear fusion converting hydrogen into helium and in the process becoming hot and bright. This is the most stable stage of their existence; it can last for about 5 billion years until their hydrogen supply gets depleted.

As the hydrogen fuel in a star gets exhausted, it begins to die and becomes a giant or super-giant, with the core contracting and the outer layer expanding. These stars eventually explode shedding their outer layer and becoming either a planetary nebula or supernova and then becoming white dwarfs, neutron stars, or black holes—all depending on their mass. Smaller stars eventually become faint white dwarfs (hot, white, dim stars) and with the depletion of nuclear fuel morph into cold, dark, black dwarfs.

All the chemical elements that we find in our universe, and here on the Earth, were created through the progressive fusion of hydrogen in to helium and helium into heavier elements inside of a star and spread into space through supernovae. Also, during a supernova, all elements heavier than iron are created when there is a release of energy plus subatomic neutron particles.

FORMATION OF SOLAR SYSTEM - Earth and Life forms

It is postulated that a whole solar system—that is, a star surrounded by orbiting planets—can be formed during the formation of a star from the gravitational collapse of a gas cloud. While the centre of a cloud collapses to form a star, the rest of matter gets flattened into a proto-planetary disc.

According to the proto-planet or nebula hypothesis, planets are formed around a new star by condensing first into a disc of molecular gas and dust, embedded within a larger molecular cloud. The condensation increases until giant heated planets are formed from the disc. And over time the gas in the disc disappears leaving planets and sometimes moons around them, along with a disc of dust and debris orbiting the star. The dust and debris can also result in the formation of asteroids and comets as well as meteoroids around the star.

Our Sun is a Main Sequence Star and it was formed in our Milky Way galaxy around 4.5 billion years ago. Our Sun has a solar system, consisting of planets and asteroids and comets that revolve around the Sun. There are as well as moons that orbit planets, like the moon around the Earth. These planets and other bodies in the solar system are postulated to have been formed through accretion from the dust grains in the solar nebula left over when the Sun was formed.

Our Earth, a terrestrial planet, is part of the four terrestrial and inner solar system planets (others being Mercury, Venus and Mars). The four outward planets (Jupiter, Saturn, Uranus and Neptune) in our solar system are not rocky but rather giant gaseous in structure such as Jupiter and Saturn.

The earliest forms of life on Earth date back to around 3.7 billion years based on fossil records. Chemical elements are postulated to have combined to form carbon-based molecules to create biological cells in hot water environments. Carbon is the sixth element and has six protons in its nucleus. A water molecule, H_2O, is composed of two atoms of the first element hydrogen fused with one atom of the eighth element oxygen. One proton of the two from the combined hydrogen atoms fuses with one of the eight protons from the oxygen atom.

Both carbon and water are crucial to life, each has special chemical properties; carbon can form stable bonds with many other elements including itself and water is a universal solvent. Over time life on the Earth evolved from carbon and water based single biological cells to multicellular organisms. Eventually the continuing evolution over billions of years produced plants and animals and apes and humans as a part of millions and millions of species of life on Earth.

The Homo Sapiens, the species to which we modern humans belong, is considered to have evolved on the Earth only around 300,000 years ago. The name Homo Sapiens comes from Latin for wise man and it was introduced only in 1758. Humans are thinking creatures, thinking being attributed to the neocortex, the outer layer of the human brain.

Along with the neocortex, responsible for language and imagination, the human brain is also composed of earlier evolutionary parts: reptilian (oldest—for controlling the body's vital functions like breathing and temperature) and limbic (mammalian—responsible for emotions and value judgment among others) brains. The three components work together through numerous interconnections to give us humans the ability to be self-aware, curious, creative and enterprising.

HUMAN ENTERPRISE FOR USHERING SPACE AGE

All ancient cultures studied the motions of our Sun and the moon but also noted and recorded constellations of stars as well as the passages of comets in our sky. Most ancient cultures were also aware of planets other than the Earth and the existence of asteroids in our solar system. Ancient humans have kept records of celestial events like eclipses of the Sun or the moon and some mapped the sky, capturing all what their naked eyes could see on a clear night. At one time, these humans postulated that our Earth was flat and at another time that the Earth was the centre around which the universe revolved.

Copernicus was the first to postulate the heliocentric view, meaning that our Sun is the centre around which the Earth orbits, and not the other way around. Galileo Galilei, an Italian astronomer, later in the same year (1692) through his observations and measurements confirmed and promoted the Copernicus model. Ever since then the Earth has been regarded as a sphere that revolves around the Sun. Moreover, in this verified model, our Sun is also the centre around which revolve all the other planets in our solar system.

Galileo was also the first to use a telescope to observe and record details about other planets in the solar system such as Jupiter's moons and Saturn's rings. Since Galileo's earliest use of the telescope to observe planets and moons, science and technology have developed other sensors including more sensitive and finely tuned telescopes to not only look within our solar system but to also observe and record our universe beyond. The two related sciences—Astronomy and Cosmology—provide a good comprehensive record of our understanding over time of the universe and its evolution.

However, it was not until 1957 that what we call the Space Age was born. It was the year when the old Soviet Union (the Union of Soviet Socialist Republics, USSR) successfully launched a satellite named Sputnik into space to orbit Earth. Sputnik operated for about three weeks at a 577 km altitude orbit and it radically changed not only how we observe our universe but also how we view and study our home planet Earth.

The first satellite in space, Sputnik demonstrated that humans can devise a rocket that can achieve sufficient acceleration upon liftoff to counter Earth's gravity and attain high enough speed and altitude to launch a satellite to orbit the Earth. This historic milestone ushered in the Space Age, the start of our access to space, meaning that humans were not confined anymore to the pull constraints of Earth's gravity.

Since Sputnik, more than 8900 satellites and spacecraft have been launched into space for a variety of purposes. Spacecraft that go around, or orbit, the Earth or other planets are artificial satellites whereas our moon and the moons of other planets are natural satellites. Many spacecraft launched by a number of countries have also ventured to and explored nearby planets, moons, comets and asteroids.

One spacecraft, Voyager 1, launched by the United States of America (USA) in 1972 is exceptionally notable. In 2012 it became the first humanmade object to travel far enough away to exit the heliosphere (the extent of our Sun's influence in space) and thus to go beyond into interstellar space, the space between stars. Voyager 1 was followed by Voyager 2 as it also achieved the same milestone of entering interstellar space in 2018.

Earth-orbiting satellites are now widely used for a variety of operational and scientific applications. These satellites are employed for many practical uses such as communications; navigation; and Earth Observation (EO) for weather and environment monitoring, and renewable and non-renewable resource mapping. These satellites have become essential space assets for the betterment of human life, an integral part of our infrastructure on the Earth.

All this historic human enterprise is just about 60 years old, but it has provided a solid foundation for the future use of space for improving our knowledge and understanding of our universe and our place in it, and perhaps more importantly for enhancing and safeguarding all life on Earth.

3. EARTH IN SPACE

Our planet Earth is part of a star's planetary system that in itself is part of a galaxy. Starting from the Big Bang, an event estimated to have occurred around 13.7 billion years ago, Space with its matter and energy is in ever continuous change and evolution that eventually produced these galaxies, stars and planets.

Over time elementary particles of matter with the help of the four fundamental forces coalesced into atoms, and atoms fusing and burning brightly into stars and stars clustering together into galaxies, resulting in millions of galaxies each containing billions of stars. Our Sun, around which the Earth revolves, is one of these stars about 4.5 billion years old in one of these galaxies named the Milky Way due to its appearance as such from Earth.

The Milky Way is a spiral galaxy in the shape of an elliptical disc with a diameter of about 100,0000-180,000 lightyears (the distance light travels in a year). It contains billions of stars and is considered to have a black hole called Sagittarius A at its centre, the same as in other galaxies. The major galaxy closest to the Milky Way at a distance of about 2.5 million lightyears is Andromeda, another spiral galaxy. These two are gravitationally bound together and are approaching each other at about 110 km per sec.

FEATURES OF SUN AND PLANETS - Heliosphere

Our Sun is situated in the outward arm of the Milky Way, about 26,000 lightyears from the galactic centre. The nearest star to the Sun is Proxima Centauri about 4.22 lightyears away, and a binary star pair Alpha Centauri A and B are close by at a distance of around 4.3 lightyears from the Sun. Our Sun orbits around the galactic centre at approximately 220 km per sec, taking about 230 million years to make one orbit. The Milky Way itself as a whole is moving at about 600 km per sec in space.

The Sun has a volume of 1.4×10^{27} m^3 and about 1.3 million Earths can fit into it. The Sun with a mass of 1.989×10^{30} kg, is about 330,000 times the mass of the Earth. About 73% of the Sun's mass consists of hydrogen and the rest is mostly helium. The Sun's mass accounts for 99.86% of the total mass of the entire solar system.

Our Sun's influence reaches far into space, a vast bubble-like region known as the heliosphere with a boundary at the heliopause beyond which the interstellar medium or space starts. The heliopause is approximately 18 billion km from the Sun, and this heliosphere is continuously inflated by plasma (hot charged particles), known as solar wind originating from the Sun.

The Sun is a sphere of hot plasma with a magnetic field. At the Sun's interior it is hot and dense as a result of a nuclear reaction through the fusion of hydrogen nuclei into helium. This reaction converts around 4,000 million kg of matter into energy per second thereby producing the light and heat that continues to expand throughout the heliosphere.

Within the heliosphere is our solar system, composed of planets as well as comets and asteroids, all orbiting counterclockwise around the Sun. Our Earth is one of the eight planets, the third nearest to the Sun. The four nearest ones (Mercury, Venus, Earth, and Mars) are terrestrial planets with rocky surfaces, while the four remaining (Jupiter, Saturn, Uranus, and Neptune) are large gaseous or icy planets. Pluto is not called a planet anymore due to its really small size; it is now called a dwarf planet.

The orbital distance from the Sun to Earth is about 149 million km and it is called 1.0 Astronomical Unit (AU). At this scale the order of planets is Mercury (0.38 AU), Venus (0.72AU), Earth (1.00 AU), Mars (1.52AU), Jupiter (5.20AU), Saturn (9.58AU), Uranus (19.14AU), and Neptune (30.20AU). The heliopause that defines the end of the heliosphere, i.e., the end of the Sun's influence in space, is at around 120 AU. The largest planet is Jupiter, about eleven times the Earth's diameter and 300 times its mass, and the smallest is Mercury that is twenty times less massive than the Earth and its diameter is around 2.5 times smaller.

There are also millions of asteroids, also called minor planets or planetoids, orbiting the Sun, the vast majority of them being in the asteroid belt located between the orbits of Mars and Jupiter. In addition, there are thought to be one trillion comet-like bodies, small snow or ice dirtballs, that have highly eccentric elliptical orbits around the Sun. So far, around 6619 comets have been discovered and they are thought to originate from the Kuiper Belt that lies beyond the orbit of Neptune or from the Oort cloud that is even farther away.

When a comet is closer to the Sun, its surface heats up and spews dust and gases producing an atmosphere around its nucleus and a long tail that can sometimes be seen with a naked eye from the Earth. There is some evidence that a massive asteroid or comet impacted the Earth and resulted in the extinction of the dinosaurs and other species about 66 million years ago. There are millions of meteoroids—a rocky or metallic body smaller than 100 m—entering the Earth's atmosphere from outer space every year. Most are fragments from comets or asteroids and are significantly small particles.

Entering Earth's atmosphere at a speed in excess of 20 km per sec a meteoroid heats up and sometimes produces a streak of light and when there are many of these entering in a series and at the same location in the sky it is called a meteor shower. Resulting from a comet, the most visible meteor shower in the year is typically the

Perseids which peaks around mid-August followed by the Leonid which peaks in mid-November. They are named after the constellations Perseus and Leo, the locations in sky from where they seem to hail. If a meteoroid survives the descent through the atmosphere and lands on the ground it is called a meteorite.

Six planets have moons that orbit them totalling more than 190 moons in the solar system. Saturn with eighty-two moons, considered to be the highest in the solar system, is followed by Jupiter that has seventy-nine moons. With a diameter of 5,262 km, Ganymede, a moon of Jupiter, is the largest moon—bigger in size than the smallest planet, Mercury.

Our moon is the fifth largest of all the moons in the solar system and has a diameter of around 3474 km. At a distance of 384,400 km from the Earth, the moon is gravitationally locked with the Earth, keeping its same face towards the Earth as it orbits the Earth. The moon reflects Sunlight making it particularly visible from the Earth during the night. The moon has an average albedo of 0.12, meaning it reflects about 12 % of radiation that falls on it. In comparison, the Earth has an albedo of 0.30, meaning it reflects more Sunlight back into space.

Both the Earth and the moon together orbit around the Sun. When the Earth is between the moon and the Sun there is a moon eclipse on the Earth as the Earth gets in the way of the Sun's light hitting the moon. When the moon is between the Earth and the Sun there is a Sun eclipse on the Earth, as the moon blocks the Sun's light hitting the Earth. Eclipses of the moon or the Sun occur about twice each during a year but some are more prominent than others.

FEATURES OF EARTH

Our Earth is around 4.54 billion years old and is composed of a number of layers. It consists of solid inner core at the centre, followed outwardly by a fluid outer core, a mantle and a crust at the outside. The thickness of the crust varies from 5 km at ocean floor to 70 km on land. The land portion is called the continental crust and it is made up of rocks consisting primarily of silica and alumina known by the term "sial". The mantle is almost 3000 km deep and is made up of slightly different silicate rock with more magnesium and iron.

The tectonic plates composing the Earth's surface are a combination of the crust and the outer mantle, also called the lithosphere. There are twelve of these large semirigid plates of irregular shapes and sizes and they move very slowly, around a few centimetres per year. When they bump against each other the boundary along where they touch is called a fault line, earthquakes can be the result.

The outer core is made up of iron and nickel and since it is very hot (4400 to 5000+ K), these metals are in liquid form. It is due to the outer core that the Earth has a magnetic field that makes a protective barrier in space around the Earth. This magnetic field shields us from the Sun's damaging solar wind.

The Earth's inner core is also made up of iron and nickel but it is solid even though it is the hottest part of the Earth at over 5000 K (about as hot as the surface of the Sun). Being deep within the Earth, it remains solid due to the tremendous pressure exerted by the upper layers.

The Earth is not totally spherical, but more like an ellipsoid. Its equatorial radius (6378.136 km) is greater than its polar radius (6356.754 km) by 21.38 km. Currently, the Earth's magnetic north pole is in the Arctic and the south pole is in Antarctica; the two magnetic poles are known to have switched positions a few times in the past. Moreover, the geographical location of the north pole is known to move around in the Arctic.

The Earth's magnetic field, i.e., its magnetosphere, extends well into space up to about 65,000 km on the day side—a result of compression from the solar wind—and on the night side it extends as a tail exceeding 6,300,000 km in length. Around the Earth, there is a zone called Van Allen radiation belt extending from an altitude of 640 km to 60 58,000 km. This zone consists of energetically charged particles—electrons and protons—thought to be from the solar wind and cosmic rays. They are trapped there by the Earth's magnetic field. The spatially extended magnetic field is thus shielding the Earth from these harmful particles.

The Earth spins around a geographical axis that is tilted at an angle of 23 degrees to its north-south magnetic axis. This east to west, or counterclockwise rotation around the axis results in the daily day/night, 24-hour cycle on the Earth. When a hemisphere is facing the Sun, it experiences the Sunlight while the other hemisphere is in darkness. The Earth revolves counterclockwise around the Sun in an elliptical orbit. It takes 365 plus days, or one year, for the Earth to complete one orbit around the Sun.

The Earth's orbit combined with its tilt results in the yearly cycle of four typical seasons—spring, summer, fall and winter. These seasons occur as the Sunlight falling on the Earth changes due to the changing distance from the Sun. As well the tilt causes one hemisphere to be closer towards the Sun during part of the orbit while the other hemisphere is away.

The Earth's northern and southern hemispheres view different parts of the night sky. From the north, forty-eight star constellations such as Orion and Ursa Major and the North pole star Polaris are prominent while from the south, Alpha Centauri A and B and the Southern Cross constellations are seen in the sky.

Earth has an estimated mass of 5.972×10^{24} kg. This mass exerts a gravitational force or pull on nearby objects imparting them with an acceleration due to a gravity of 9.8 m per sec per sec. This gravity force experienced by a body near Earth decreases as its distance from the centre of Earth increases.

The lowest velocity that a body must have in order to escape the gravitational attraction of a particular celestial body is called escape velocity. The escape velocity required from the surface of Earth is 11.186 km per sec, meaning a spacecraft must be going at that speed outward without ever falling back to the surface of the Earth. For Mars it is about 5.03 km per sec as Mars is about one third in size and mass compared to Earth. Whereas for moon it is around 2.38 km per sec. The escape velocity from the surface of the Sun is 618 km per sec.

Chemical Elements on Earth

Chemical elements constitute all the matter of the universe and they are also the building blocks of all the material and living organisms found on Earth. There are known to be 118 chemical elements on Earth, out of which ninety-four occur naturally, the remainder are synthetic or made by humans. Each element is unique and based on the number of protons in its nucleus.

All the universe's ninety-four naturally occurring elements are postulated to have been produced through cosmic processes; the first few elements (hydrogen and helium) were created in the beginning of the Big Bang, while others up to the element iron were created through nuclear fusion in stars and the subsequent heaviest elements were produced through nuclear synthesis in supernovas, the explosive death of stars.

The elements are arranged in a sequence in rows and columns in what is called the Periodic Table with increasing Atomic Number 1 to 118. The Atomic number is equal to the number of protons in the nucleus, the first element being hydrogen with one proton and the last being oganesson (a synthetic element) with 118 protons, and both are gases. The number of protons plus the number of neutrons constitute an element's mass number. A period in the Periodic Table is a row of elements having the same number of electron shells. And each column is a group of elements having similar chemical and physical properties. Together the rows and columns depict the periodic law of chemical elements.

The number of neutrons for a given element, however, can vary, with such different atoms being called isotopes. Some isotopes have unstable nuclei, resulting in radioactivity through the release of subatomic particles in their effort to become more stable, giving off energy-radiation-in the process. For example, carbon, the sixth element has six protons and six neutrons and thus gets labelled as carbon-12. However, a small amount of carbon exists on Earth as radioactive carbon-14 (six protons plus eight neutrons), and it is the amount of carbon-14 found in a fossil that allows us to determine its age.

Oxygen the fourth element is the most abundant element in Earth's crust making up 46% of Earth's mass. Silicon, the fourteenth element, is the second most abundant at 27.7 % followed by aluminum (8.1 %), iron (5 %), calcium (3.6%), sodium (2.8 %) and potassium (2.6%).

Two atoms of hydrogen and one atom of oxygen bond to create H_2O, what we call water, the substance essential for life on Earth. Water is abundant on Earth with 71% of Earth's surface being covered with it. Water exists in three phases: liquid, gas (e.g., steam) or frozen (ice and snow). The ocean holds around 96.5% of Earth's water, the remainder being in rivers, lakes, and frozen in ice caps and glaciers. Water also exists in air as water vapour, in the ground as soil moisture and in aquifers, and in bodies of all living organisms (plants, animals and humans).

Living Organisms on Earth

A living organism is considered to be any individual entity that propagates the properties of life (such as reproduction, growth and development, regulation, response to stimuli) on Earth. The start of living organisms on Earth can be dated back to almost 4 billion year ago; all such organisms are composed of organic matter, i.e., carbon-based molecules.

Life on Earth is carbon-based; carbon being the special element that can easily combine with other elements such as hydrogen, oxygen, and nitrogen to form complex molecules or compounds, thereby producing the basic unit of life that is called a cell. The nature of life itself, whether it is a force, energy, or phenomenon that makes a cell alive remains elusive as ever.

A cell is the smallest (ranging in size between 1-10 μm) and the simplest living organism. A cell is a building block of all living organisms on Earth. Organisms can be single-celled (e.g., bacteria) or multi-celled (e.g., fungi, plants and animals), and cells arise only through a division of a previous existing cell. A cell has a property of being able to self-divide and multiply, replicating itself, that is it is an entity or life form, that propagates the properties of life. As such, viruses are not considered alive as they cannot replicate on their own but can only do so in a host living cell.

All cells are mostly composed of water (at least 70% of a cell's mass) and carbon-containing or organic molecules. Cells have three major features: a nucleoid or nucleus (central portion containing genetic material), cytoplasm (semifluid matrix or gel filling the interior), and a plasma membrane (phospholipid bilayer embedded with proteins surrounding the cell). The membrane separates a cell from its surrounding environment, and it is water that transfers or carries nutrients into and waste out of a cell; water is essential to sustain life.

There are two main classes of cells, defined by whether they contain a nucleus. Prokaryote cells produce the simplest organisms such as bacteria which are single-celled and do not have a true or distinct nucleus. Eukaryotes cells with a distinct nucleus can produce single- or multi-celled organisms and are more complex than prokaryote cells and form the basis for virtually all life—plants, animals and humans—that we see today. In eukaryotes cells, the DNA (deoxyribonucleic acid—a molecule in the form of a double helix that carries genetic instructions) is linear and well organized inside a nucleus.

All living organisms depend on a source of energy (the ability to perform work or function) to survive, and cells have chemicals that store and release the energy to drive reactions in each cell. For example, plants use light energy from the Sun to produce chemical energy and ultimately its structural components through the process called photosynthesis that converts light and carbon dioxide and water into rich organic compounds and in so doing releases oxygen into the air. Other organisms such as animals obtain energy by ingesting food sources (e.g., plants and animals) and chemical energy from the environment.

All living organisms must also be capable of releasing energy stored in food molecules through a chemical process known as cellular respiration. In aerobic respiration, air that contains oxygen is breathed in while carbon dioxide is breathed out. Virtually all eukaryotic organisms perform aerobic respiration.

Humans and other animals must respire aerobically, breathing air in and out as they require oxygen for survival. Most marine living organisms extract oxygen from water while there are some like phytoplankton (plants) who use photosynthesis for their survival and release oxygen in the process. Thus, plants are crucial to life on Earth that requires oxygen to breath.

Single-celled life is thought to have originated around 4 billion years ago on Earth. The evolution of photosynthesis in the ocean for producing oxygen is considered to have begun around 3 billion years ago. But it was only about 600 million years ago that first multi-celled animals such sponges and jelly fish emerged in the ocean. By 470 million years ago the first plants grew on land with the first land animals such as millipede appearing around 428 million years ago.

It is only after dinosaurs went extinct about 65 million years ago that mammals rose in their ecological prominence. Mammals are vertebrate animals constituting the class Mammalia and they possess neocortex brains and their females have mammary glands that produce milk for feeding young. There are more than 5450 species of mammals that include humans as well as primates such as monkeys and apes.

A species is a group of living organisms that has similar individuals capable of interbreeding and passing on their genetic material to their offspring. To date more than 1.3 million species currently on Earth have been identified and described. However, it is estimated that about 8.7 million species live on Earth with around 6.5 million species on land and 2.2 million in the ocean. The nature of life itself and the consciousness imparted to a brain to make it a mind in living organisms remains beyond scientific investigations.

Humanity's March

It is estimated that about 107 billion humans have ever lived on Earth through the last 200,000 years or so since humans first appeared on Earth. There are currently around 7 billion humans alive today.

It is worth noting that in the 13.8 billion years evolution of our universe, the span of recorded human history is roughly only 5000 years old. Industrial revolution brought on by humans started about 200 years ago, in the 18th century, when the transition to new manufacturing processes began. The petroleum using automobile cars arrived not too far back in 1879.

Digital electronic calculating machines were developed during World War II and semiconductor transistors arrived in the late 1940s. The integrated circuit chips were first produced in late 1950s and that led to the subsequent digital revolution and the age of the personal computer. All this innovation contributed to the start of the Space Age with the launch of Sputnik in 1957, a historical milestone that is only less than 70 years old.

In just a short span of about 100 years, we humans have indelibly changed the face of Earth and its environment. It is an astonishing reflection of human nature and a cause for caution for the future of our species and for all life on Earth.

4. NEAR EARTH SPACE ENVIRONMENT

Earth is surrounded by an atmosphere, an envelope of gases that we call air, which interacts with the energy coming from the Sun and the radiation arriving from outside the solar system. This gaseous envelope is kept in place by Earth's gravity. The atmosphere is crucial for the protection and sustainability of life on the Earth, as these gases absorb ultraviolet and other radiation that is harmful to life.

Importantly, the atmosphere also warms Earth's surface through heat retention thereby reducing temperature extremes during day and night and also tempering the day to day variations. All this moderation results in temperatures that allow liquid water to exist on Earth's surface.

EARTH'S ATMOSPHERE

The atmosphere of Earth is composed of many layers, and atmosphere's pressure and density of constituents decrease with increasing altitude from the surface of Earth. The bottom layer starting from the surface to about 12 km in altitude is called Troposphere. This is followed by Stratosphere until about 50 km in altitude, then Mesosphere until 80 km in altitude and Thermosphere until 400 km in altitude. Finally, at the top, the outermost layer is called Exosphere and that continues up to about 100,00 km, making it the upper limit of Earth's atmosphere.

Within these layers is Ionosphere, that stretches from about 48 km to about 1000 km in altitude and it is a critical layer in the chain of Sun-Earth interactions. The Ionosphere is a dynamic region that grows and shrinks based on solar conditions and sub-divides further into D, E and F sub-regions; based on what wavelengths of solar radiation is absorbed therein.

The Ionosphere is a layer of electrons and ionized atoms and molecules and it is what makes radio communications on Earth possible. Radio waves transmitted anywhere from Earth's surface, propagate outward and get reflected back from the Ionospheric layers towards Earth, thereby allowing the radio signals to travel along Earth's curvature.

The atmosphere of Earth is dry air and water vapour. Dry air is primarily composed of nitrogen (N_2 -78%) and oxygen (O_2-21%) with some argon and other trace gases including about 0.04% carbon dioxide (CO_2). Water vapour in the atmosphere is about 0.25%.

Ozone (O_3) in the upper reaches of the atmosphere blocks the Sun's harmful-to-life ultraviolet radiation reaching the Earth's surface. Moreover, other constituents of the atmosphere also have some impact on the spectrum of the Sun's radiation passing through it. This is known as absorption spectrum and it is used in measuring the concentrations of atmospheric constituents.

Carbon dioxide (CO_2), methane (CH_4) and other carbon-based gases cause the heat generated at Earth's surface to be trapped resulting in what is known as the greenhouse effect. This is the same phenomena that we see in a greenhouse built of transparent material like glass that allows solar radiation or light to come inside but not allowing heat that is of longer wavelengths to escape from it, thereby raising the inside temperature. Greenhouse gases such as carbon dioxide (CO_2), methane (CH_4), nitrous oxide (N_2O), and ozone (O_3) as well as water vapour (H_2O) absorb and emit radiant energy within the thermal infrared range.

The concentration of carbon dioxide in Earth's atmosphere is steadily increasing over the last hundred years and this increase is attributed to human activity, principally the burning of fossil fuel.

As a result, the average temperature at the surface of the Earth is going up, making the last ten years the warmest in recorded history.

Global warming due to human activity is seen to be changing Earth's climate. Even a less than 2 K rise in global temperature is considered catastrophic for life on Earth. This small change is projected to alter normal weather patterns and produce extreme events such as intense heat waves and wild fires, draughts, heavy precipitation, stronger hurricanes and extensive floods.

The warming of the Earth results in melting of sea ice and glaciers that would raise level of water in oceans. The rising sea levels will cause devastating effect on coastal habitats and destructive impact further inland. Climate Change is for real and it impacts ecosystems, agriculture, biodiversity, and economy and ultimately life on Earth for the worse.

OUTER SPACE

Earth's atmosphere at upper altitudes is very sparse. For example, 99% of the atmospheric mass is concentrated in the first 32 km. At an altitude of 100 km the atmospheric pressure is just one millionth of the value at sea level. As such, a distance of 100 km above Earth is thought to be the limit for the functioning of an aircraft as the air above becomes too thin to support an aeronautical flight.

At altitudes above 100 km, a vehicle has to travel really fast, at an orbital speed to have the aerodynamic lift to support itself. Hence, for legal and regulatory purposes the boundary between atmosphere and Outer Space gets defined at an altitude of 100 km above Earth. Any vehicle flying in Outer Space is called a spacecraft and it gets regulated as such.

Outer Space is a harsh environment for the human body, due to the lack of protection that is provided by Earth's atmosphere and magnetic field. The space environment is composed of high energy radiation and particles from our Sun and from beyond our solar system, like cosmic rays. Some of these particles such as electrons get trapped in the Earth's magnetic field thereby producing Aurora

Borealis in the northern and southern hemispheres around the two magnetic poles.

The near-Earth space thermal environment has more than 220 K temperature differential from a high of 375 K when facing the Sun to a low of 173 K when shaded or in the dark. Thus, for an object to survive or function in outer space it has to deal with this extreme temperature variation and exposure to the ever-present harmful radiation.

5. TRAJECTORIES IN SPACE

The motion of a spacecraft and its location in space is governed by a set of laws of physics and these laws describe its trajectory or orbit in a predictable and precise manner. These laws of universal gravitation and of motion were formulated by Newton and they laid the foundation for space exploration. These laws define what forces are exerted in space and how a body in space moves and how its motion can be changed or altered.

LAWS OF GRAVITATION AND MOTION IN SPACE

Newton's law of universal gravitation states that every particle attracts every other particle in the universe with a force that is directly proportional to the product of their masses and inversely proportional to the square of the distance between their centres. This means that a massive body such as the Sun or Earth exerts a force, called gravity, to attract other bodies around it. This exertive influence decreases significantly as the distance away from this massive body increases.

Newton's First Law of motion states that a body continues to be in a state of rest or motion unless and until an external force is applied to it. This means that a body in space such as a spacecraft will continue to move at its attained speed unless impacted by some force. There are essentially only two external forces in space: the pull of gravity of nearby celestial bodies or the drag caused by air resistance when passing through any atmosphere surrounding these bodies.

Newton's Second Law states that a body affected by a net force will accelerate in the direction of the net force. This acceleration, or change in speed, is directly proportional to the magnitude of the net force and inversely proportional to the mass. An external force such as that of gravity produces acceleration while air resistance produces de-acceleration. As such, an object being subjected to gravity attains speed and an object experiencing atmospheric drag loses speed.

Newton's Third Law states that every action produces an equal reaction in the opposite direction. Thus, a body such as a rocket or a spacecraft that expels exhaust at one end produces a thrust or force in the opposite direction thereby creating an acceleration of that body in that direction.

As such Earth's force of gravity resulting from its mass imparts an acceleration of 9.8 m per sec per sec to objects near its surface. This acceleration due to gravity is experienced by any object near Earth, and is the same, independent of its mass. However, the impact of gravity reduces as the away distance increases, meaning Earth's influence on satellites in near-Earth orbits is around 90% of what it is at its surface.

Acting as a centripetal force, the gravitational pull draws a satellite into a free fall towards Earth's centre. This makes a satellite to go around and around following Earth's curvature, and eventually to enter Earth's atmosphere and burn-out from air friction.

PLANETARY ORBITAL MECHANICS - Kepler's Laws

While Newton's laws provide important theoretical foundation, it was however, Johannes Kepler, a German astronomer, in 1605 who studied the empirical observations of planets moving around the Sun. Thus, he was the first to deduce and provide mathematical formulations for the motion of planets around the Sun based on these empirical observations. These were captured in Kepler's three laws of planetary motion or motions of bodies in space around other heavier or more massive bodies.

Kepler's First Law is known as the Ellipse Law as it states that the path of a planet is an ellipse with the Sun being at one of its fixed point- a focus. This means that the orbit of a smaller body in space around a bigger, more massive body is described by an ellipse, an elongated sphere, like an oval. For planets, the Sun is the dominant body, e.g., the mass of Jupiter, the most massive planet is only a thousandth that of the Sun. Planetary orbits around the Sun are ellipses and all are in the same plane.

Kepler's Second Law is known as the Area Law. In the Area Law he deduced that the planetary elliptical orbit subtends equal area in equal time, meaning that planets nearer the Sun move faster than when they are farther away.

Kepler's Third Law is known as the Harmonic Law; the square of the period of a planet (time it takes to make one complete orbit) is proportional to the cube of its mean distance to the Sun. This means for example that as planet Mars is 52% farther from the Sun compared to Earth, its period or Martian year is 88% longer. There are 687 days per year for Mars and 365 days per year for Earth.

As per Newton's laws and Kepler's laws derived from observations, a planet or a satellite stays in its orbit due to the balance between the centrifugal force (radially outward from the axis of rotation) resulting from its speed and the force of attraction, the gravity. All this means that a planet needs to move faster nearer the Sun to counter balance the added pull from the Sun.

For example, Earth at a distance of 1 AU moves around the Sun at a speed of 29.3 km per sec while Mercury at 0.4 AU being closer to the Sun moves 60% faster at 47.3 km per sec. Mars being farther from the Sun at 1.5 AU moves 20% slower than Earth at 24 km per sec whereas Neptune, the farthest planet at 30 AU has a speed of 5.4 km per sec.

The position of a planet or a satellite in its orbit is described by its ephemeris, a collection of six parameters known as Keplerian elements: length of the semi-major axis and eccentricity of the ellipse plus inclination with respect to the elliptical plane with respect to the reference plane, position of the perihelion (closet point to the Sun) or perigee (the closet point to the Earth for Earth-orbiting satellites), time since perihelion or perigee, and position of the node (ascending node for Earth-orbiting satellites from the south to the north pole–it is descending node the other way around).

These mathematical formulations are essential for determining and predicting orbits and trajectories of objects in space and how much force or thrust would be required to alter them. In space voyages, there are two interrelated terms employed called orbital and escape velocities.

The orbital velocity is the speed required to achieve orbit around a celestial body, such as a planet or a star, while escape velocity is the speed required to leave that body's gravitational pull. The escape velocity is around 41% faster than orbital velocity. The escape velocity from Earth's surface is about 11.2 km per sec, while that from the moon is 2.38 km per sec and from Mars is around 5 km per sec.

SATELLITE ORBITS AROUND EARTH

For satellites orbiting around Earth, the Earth is the dominant or the massive body. Even the moon has only one thousandth of Earth's mass. A satellite at an altitude of 100 km above Earth, where Outer Space begins, needs to have a minimum orbital velocity of 7.84 km per sec to attain a stable orbit.

The required orbital velocity decreases with increasing altitudes. For example, the moon travels only at a velocity of 2 km per sec at a distance of 300,000 km from Earth. While satellites at higher altitudes move at slower speeds, their orbital periods (time to make one orbit around Earth) are longer.

Earth-orbiting satellites are normally classified based on the altitude and inclination of the orbit plane to Earth's equator.

- Three altitude ranges around the Earth are important for a satellite's orbit: up to 2000 km is called a Low Earth Orbit

(LEO), from 2000 km to 30,000 km is called a Medium Earth Orbit (MEO), and an altitude of 35,780 km is known as a Geosynchronous orbit. A Geosynchronous orbit has an orbital period of 24 hr thereby matching Earth's rotation on its axis. Almost all satellite orbits used for applications on Earth are near-circular.

- An inclination of 0 degrees to the equator gives an equatorial orbit while an inclination of around 90 degrees provides a polar orbit; equatorial orbits have their ground tracks along the equator while polar orbits have north-south ground tracks.
- A Geosynchronous orbit with 0-degree inclination is special. In this orbit at around 36,000 km in altitude, a satellite travels at the same speed as the turning or rotation of the Earth. This synchronicity makes a satellite appear stationary or fixed over a particular location along Earth's equator. This Geostationary orbit thus is particularly useful for communications and weather watch applications.
- Highly elliptical orbits can also be of use and one such is called Molniya orbit. These orbits make use of the property that a satellite travels faster near a perigee and slower at apogee, thus making a satellite to be seen as almost hovering over the same location on Earth. Thus, a Molniya orbit with a proper degree of inclination can be spending most of its orbital time over a particular geographical location and thus is seen to be quite useful for communications satellites for northern latitudes.
- LEO orbits are predominantly used for looking down on Earth, and have a period of about 100 min, meaning there are typically about fifteen orbits around the Earth in 24 hr.
- LEO orbits can be Sun synchronous, where the orbital plane remains at a fixed angle towards the Sun, i.e., the precession of the orbital plane or its rotation at 0.9865 degrees eastward each day to match the Earth's movement around the Sun. This makes a satellite to pass over the equator at the same local time on its ascending node. A Sun-synchronous orbit provides

consistent solar illumination throughout a year and is particularly useful for those satellites that employ optical sensors.

- LEO satellites in polar orbits provide nominal global coverage of the Earth. In a polar orbit a satellite travels north-south while the Earth is rotating east-west. This makes a satellite orbit to observe or have access to practically all areas on the Earth.
- Orbital inclinations of less than 90 degrees, provide for direct or prograde satellite orbits where their successive ground tracks typically move from west to east crossing the equator. And orbital inclinations of more than 90 degrees provide for retrograde satellite orbits where their ground tracks move from east to west. These LEO orbits usually display a 3-day, 5-day or 7-day repeat sub-cycles as ground tracks transverse around the globe and revisit the same location on the Earth.
- Given the diversity of satellite orbits around Earth it is important to select a proper altitude, inclination and synchronicity to suit a particular application. This selection is made on the basis of an analysis of trade-offs to find the best ground track and revisit coverage that can be attained throughout the duration of a mission.

There are particular locations in space that result in distinct satellite orbits or spacecraft trajectories. Particularly notable are locations in space between two massive bodies where their competing gravity pulls from opposite directions negate each other. At such a location a small body like a spacecraft appears to be locked into the motion of the two larger bodies. Such locations are called Lagrange points where the combined gravitational forces of the two large bodies create a point of equilibrium where a spacecraft appears to have been parked.

There are Lagrange points between the Earth and the Sun and between the Earth and the moon. A satellite placed at one of these points maintains its position relative to the larger bodies, making it useful for space explorations, space weather observations and other applications.

SATELLITE ORBITAL DYNAMICS

Typically, a satellite atop a rocket is lifted from the ground with the ignition of the chemical fuel stored in the rocket. This burning of fuel produces a downward exhaust that results in the upward thrust or lift of the rocket. The fuel-burn and the associated exhaust is controlled to produce the tremendous force required to raise a heavy rocket containing all that fuel plus the satellite sitting on the top. It is necessary to keep increasing a rocket's speed to counter Earth's gravity, i.e., its downward acceleration or pull.

It is usual for a rocket to have two or more stages of fuel storage and burn, each stage gets separated and discarded when the fuel in it is depleted. It is the final stage of a rocket that releases a satellite into its intended orbit, after that stage has attained the required altitude and speed. Once deployed into its orbit, a satellite must rely upon its own resources, the job of the rocket is done.

- The kind of orbit attained by a satellite (altitude and inclination) initially gets decided by the speed, altitude and location of the rocket at the time the satellite is ejected or deployed from it. A satellite post separation is in non-powered flight at its deployed speed and is thus in a state of free fall towards the centre of Earth.
- If a satellite has onboard fuel and a thruster it can adjust its orbit after deployment. However, there are many satellites that do not carry any fuel and thus have no means of raising their altitude. A satellite's altitude continues to decrease over time due to the atmospheric drag.
- The thrust required to change a spacecraft trajectory or satellite orbit is usually stated through the change required in its speed, a parameter called delta-V. The delta-V required first to reach Low Earth Orbit starts at 9.4 km per sec given that atmospheric and gravity drag associated with launch typically adds 1.3-1.8 km per sec to the launch vehicle delta-V required to reach normal LEO orbital velocity of around 7.8 km per sec.

- The minimum delta-V needed to go from a LEO to a GEO orbit is about 4 km per sec. Once in orbit, the delta-V required for compensating atmospheric drag in LEO can be around 7-100 m per sec per year while in GEO the station-keeping can take around 50 m per sec per year.
- A mission typically makes a delta-V budget as a part of its mission design based on the propulsive manoeuvres required to be performed during the mission life. And after each manoeuvre a tally is kept of how much fuel is left and the associated delta-V that can be imparted.
- Delta-V is typically applied at two particular locations, perigee and apogee, in a satellite's orbit that will result in the intended change. A thrust applied at perigee, the closet point to the Earth, increases a satellite's apogee, the farthest point from the Earth and vice versa, i.e., a thrust at apogee raises perigee.
- A repeated application of delta-V at perigee, increases apogee, thereby increasingly raising a satellite's altitude above the Earth. This increasing of apogee approach is used for a spacecraft launched on an interplanetary voyage.
- To circularize an orbit, delta-V is applied at apogee so as to raise the perigee. Also, delta-V applied at apogee can be used to change the inclination of an orbit.
- For launching a satellite, a rocket needs a large force for the propulsion. Typically, liquid chemical fuels are used as propellants such as kerosene, alcohol, liquid hydrogen, or hydrazine (N_2H_4-inorganic chemical compound that is toxic, flammable liquid).
- A launching rocket or a launcher is essentially a container of fuel that converts chemical energy into kinetic energy (i.e., it raises the speed from zero to that required for the orbit) and potential energy (to raise altitude from zero to that required for the deployment).
- A launcher can only put a small percentage (less than 3%) of its launch mass (i.e., satellite including the payload) into orbit. More than 97% of a rocket's mass is devoted for the fuel to be

consumed to overcome gravity and to increase a rocket's speed upon liftoff.

- Launch sites near equator are efficient for launching satellites into low inclination orbit to take advantage of the Earth's rotation speed. Thus, a satellite gets a boost of up to 0.43 km per sec when launched from an equatorial location.
- In-orbit propulsion is usually provided through a chemical fuel (e.g., hydrazine). For satellites operating in low Earth orbits, a delta-V is applied periodically to make up for the decrease in altitude from atmospheric drag. For interplanetary voyages, other means such as electrical propulsion or atomic fuel are used to apply delta-V and change a spacecraft's trajectory.

For satellites destined for high orbits or spacecraft travelling to escape Earth's gravity, these are deployed from launching rockets typically into relatively low Earth orbits. Some heavy launchers can deploy a geostationary satellite directly into a geostationary transfer orbit that is highly elliptical with a low perigee but an apogee at the geosynchronous altitude.

Post deployment, a satellite uses its onboard fuel to either circularize its orbit, i.e., to raise its perigee or attain the higher required orbit. This is done by increasing speed through the injection of delta-V gradually and repeatedly at perigee or apogee to escape from the existing orbit.

Eventually a spacecraft voyaging to the moon can have high enough orbit to get captured by moon's gravity. Or a spacecraft travelling beyond the moon can continue to raise its orbit to escape Earth's gravitational influence and embark on an inter-planetary voyage. Since delta-V manoeuvres require burning of onboard fuel, and this fuel is of limited quantity, judicious choice is made when to use this mission-limiting resource.

6. SPACE MISIONS

Space Missions are undertaken to meet an objective or serve a purpose. The objective can be in the area of research and development, scientific investigation, exploration, commercial exploitation, operational service, or a combination of these. Often space missions are undertaken with multiple objectives and can employ more than one satellite to fulfill mission objectives. As such, missions can have a constellation of satellites flying in synch with each other in a tandem or some other formation.

As an example of mission objectives, Canada's SciSat 1 satellite, launched in 2003, had the primary mission of measuring distribution of ozone in the Stratosphere above the Arctic. The mission requirements thus called for a space instrument that in orbit would look back at the Sun through Earth's atmosphere—an optical sensor called a Fourier Transform infrared spectrometer. This type of sensor measures Sunlight in different parts of the frequency spectrum.

The spectrum of Sunlight passing through the atmosphere is absorbed differently by different atmospheric gas constituents, thereby allowing the spectra measured by the instrument to not only provide the concentration of ozone but also of other gases. As a result, SciSat 1 mission still continues to routinely provide to the international scientific community the varying concentrations of atmospheric chemicals at stratospheric altitudes.

MISSION COMPONENTS - Space and Ground Segments

A space mission is usually concerned first with the fabrication of a space segment and a ground segment and with the development of applications or users that the mission is required to serve. Once built and tested and subsequently launched into space, it is through the in-orbit operations of a satellite over the duration of the mission that the return on investment is realized and the mission objectives are accomplished.

A space segment is usually comprised of a satellite or spacecraft (or more than one) with a launch vehicle for launching it into space. A ground segment is typically composed of an infrastructure required on the Earth to operate the satellite in space and to utilize what the satellite is designed to produce.

A satellite is usually composed of two components: a bus to perform common tasks and one or more payloads for mission-specific tasks. A bus performs basic functions that any satellite must undertake such as to communicate with the ground infrastructure, maintain the satellite's orbit and orientation, generate power for the onboard equipment and maintain satellite's structural and thermal stability.

It is a payload, however, that performs the functions for which the mission in space is undertaken. Thus, a payload is mission specific and usually unique and is the reason why a satellite is put into space.

For example, a payload can be a scientific instrument, such as an optical or radio telescope, an imager, a spectroscope, a sensor, or a communications equipment. A payload is chosen in order to meet user needs or furnish a service expected by the mission requirements.

Satellites usually are classified based on mass as it is the mass that usually determines the cost of a launch into space. Based on the kind of orbit required, launch costs can vary from $1000-10000 per kg. Efforts continue to be made to reduce these costs and make access to space more affordable.

A satellite with a mass more than 1000 kg is termed large while one between 500-1000 kg is medium-sized. A satellite less than 500 kg is called small; subdivided further into 100-500 kg as mini, 10-100 kg as micro, 1-10 kg as nano and less than 1 kg as pico.

Large satellites such as launched for communications or Earth Observations that employ many different instruments or sensors can be more than 5,000 kg. The heaviest or largest satellite named TerreStar 1 weighed at around 7,000 kg. It is a telecommunications satellite launched in 2009 by ESA's Ariane 5 launcher for operations in a geostationary orbit.

Many large and heavy satellites such as for communications purposes require stand-alone or dedicated launch vehicles. Some rockets can put into space a number of satellites in one launch and even into different orbits.

Rocket launchers come with different capacities; a launcher having a lift capacity (around 3% is for a satellite and the rest is for the fuel it must carry) of 10,000 kg is considered small whereas one capable of lifting between 20,000 and 50,000 kg is called heavy.

With a lift capacity of 140,000 kg, Saturn V was the heaviest, tallest and most powerful rocket ever developed. It was used by NASA between 1967 and 1973 for the Apollo space program for human explorations missions to the moon.

MISSION CYCLE

Development Phases

Practically all space missions follow a cycle; from conception, to design and development, fabrication or build, launch and early orbit, to in-orbit operations (called useful mission life), and finally disposal at the end of life.

The mission cycle usually gets planned and executed in phases; from A to F. Usually these phases are run in sequence, to allow for an orderly and systematic approach to understand and reduce risks. Some enterprises especially in the commercial sector can deem to accelerate this process of design, development and test by undertaking some phases concurrently or foregoing them completely.

Phase A is typically used for defining the mission in terms of objectives or purpose with some basic requirements from users and some preliminary cost estimates. It can take from a few months to a year or more depending on the scope and complexity.

Phase B is usually devoted to developing detailed requirements mostly from a user's perspective and transposing these into engineering and scientific specifications that can in fact be verified during the course of the mission. A preliminary design is also produced during Phase B with a more firmed up estimate of cost and schedule along with an identification of challenges and risks and a way around them. Phase B can take a year or two.

Phases C and D can be undertaken together as they often have an overlap. Phase C is devoted to detailed design with firmer cost estimates and a more robust identification of risks and ways of mitigating these. Phase D is known as the build phase during which fabrication takes place according to the design and smaller components are assembled together to produce the whole satellite.

During Phase D, components and the whole satellite are thoroughly tested to ensure not only that they will function according to the specifications but also that they will survive the rigours of the launch and the harsh conditions of space. These tests include what is commonly called shake and bake: subjecting a satellite to vibrations to simulate launch conditions and to heat cycles to simulate thermal conditions in orbit. Phases C and D can last 2 to 4 years or several years more depending on the challenges faced during building and testing.

Successful ground testing is followed by Launch and Early Orbit Phase (LEOP) during which a satellite is mated with the launch vehicle within the faring at the top. A satellite is fuelled at this stage and weighed and the rocket is subsequently fired for lift off and for the deployment of the satellite in space.

Satellite Launch and Operations

A satellite is deployed when the uppermost stage of the rocket reaches the desired speed and altitude. The deployment is usually activated through a spring-loaded mechanism and it can involve firing of attaching bolts. Once a satellite is uncoupled and free of the launcher, it starts to self-activate. It is then ready to interact with commands from mission control and send telemetry to ground indicating the status of its health and its orientation or attitude.

During LEOP, a process that can last for several days, all effort is made to ensure that a satellite is safe and sound, i.e., it is power positive, is within its thermal limits, is in a stable, dynamic state with the required orientation. It is often now that large appendages such as solar arrays and antennas, that were in the stowed configuration during launch so that they would fit into the fairing, are unfurled or deployed in space.

Also, as a part of LEOP, a satellite is also put into its operating orbit if it is different from its released orbit. This is achieved through the firing of thrusters using the onboard fuel to inject additional velocity or delta-V thereby raising the orbit altitude.

Following LEOP a satellite is prepared gradually to take on the task for which it was designed, built and launched. This period of commissioning can last several months during which time a satellite is subjected to tests to characterize its in-space or in-orbit performance.

During the commissioning period, the performance of the payload is calibrated. This is to ensure that the measurements that its sensor makes over time can be compared with each other and any shift in a sensor's performance can be compensated. A sensor is calibrated routinely during the course of the mission. Often external means such as ground targets are used for calibration purposes in addition to following an internal sensor calibration process.

It is after a satellite is determined to be ready or be fit for routine operations that Phase E, the operational phase starts. Phase E can last from a few months to several years in accordance with mission requirements. It is common to try to increase the duration of the in-space or on-orbit operations phase.

It is during the operations period that the benefits or results of the mission are achieved. These are the returns, usually in the form of science data, operational service, or revenues from the earlier large investments in time, budgets, human labour and other resources. Costs associated with Phases A to D are normally tremendous and the risks of failures during LEOP are high.

Phase E costs are typically a fraction, usually less than 10%, of the overall mission costs, that can be a few million to hundreds of million and sometimes even more than a billion dollars. The main reason for the high cost is that each space mission, being typically one of a kind, needs to reduce risk of failure. Often new technology needs to be developed through modelling and repeated testing and that is an expensive and long process.

Longevity of Phase E, the operations period, is dependent on many factors such as the redundancy built into a satellite and prudent use of onboard resources; particularly power and fuel. As well a catastrophic loss of some life-limiting functionality needs to be avoided. This can be a result of some hardware component or software failure due to harsh conditions of space such as radiation and thermal variations.

It is often the knowledge, skill and tenacity of an operations team on ground that determines not only the longevity of a space mission but also how well its objectives are achieved. Ground operations teams can overcome catastrophic anomalies and malfunctions to produce results far beyond the nominal mission requirements or life expectancy. For example, Canadian SciSat 1 mission although designed for a two- year mission continues to provide measurements of ozone and other chemical constituents even after 17 years of operations.

Satellite Disposal

At the end of Phase E, a satellite is prepared for its disposal in Phase F. However, due to equipment failure, it is often that a satellite becomes un-responsive to the extent that it cannot be prepared for a proper disposal.

Recent international guidelines stipulate that a non-operational satellite must be capable of entering Earth's atmosphere in less than 25 years. Further, upon reentry it must disintegrate and burn in atmosphere and thereby produce no debris to fall back onto the Earth. It is incumbent on the operator or owner of a satellite to decommission the satellite for disposal after its useful life.

Decommissioning means that a satellite is put into a state whereby it cannot create more debris in space. To achieve this state requires passivation of a satellite through a discharge of its battery and an emptying of its fuel. This is to ensure that the satellite does not blow-up and disintegrate into smaller pieces. Ever-increasing space debris is becoming a major threat to space assets. The operational means, such as raising a satellite's orbit, used to avoid debris are inefficient and costly to missions.

All space missions typically undergo milestone-based reviews over the course of its mission. These checks are often formal and externally led, to ensure a mission is on track in terms of cost, schedule and achievement of its objectives. This is required because any success of a space mission is dependent on team work. Usually a space team is composed of hundreds of experts who are involved in diverse set of activities as a mission progresses from conception, design, development, fabrication, testing and space deployment to operations and then ending with a satellite's disposal.

As such, mission reviews allow a formal process through which the whole team and all the stakeholders are kept informed of the progress of the mission. This periodic sharing of the information is conducive to creating a common understanding of any risks to the mission and the measures undertaken for their mitigation. This methodical approach leads to better and more successful mission outcomes.

7. TECHNOLOGY FOR SPACE

Space Missions generally employ technology that is specially designed, developed, built and tested to be able to survive and perform the required functions in the harsh environment of space. This level of care and effort is required so as to minimize risks and ensure success given that a space endeavour is inherently expensive (in terms of cost and human effort) and its survival is always uncertain.

QUALIFICATIONS FOR SPACE

Space is a harsh environment due to the presence of solar and cosmic radiations and charged particles and of large thermal variations that can easily cause harm to electronics, mechanical devices and equipment. Moreover, access to space requires use of powerful launching rockets capable of burning massive amounts of fuel quickly in a controlled manner and these launchers produce significant vibrations that all satellite components must be able to withstand.

All this explosive power is needed to provide the required thrust for liftoff and acceleration of a rocket and a spacecraft atop it to counter gravity's downward pull. Even with so many years of experience in designing and building successful rockets, space launch remains inherently a risky venture.

Thus, space missions require research and development of materials and design and manufacturing processes to create the needed specialized technologies that are to some extent unique. This is a long and expensive development process and only a few government agencies, companies and universities have the necessary resources and capacity, strategic vision, and commitment to be engaged in space missions and associated technologies.

Space technologies can be grouped commonly into two segments: space and ground. The former is what is required in space (satellite or spacecraft bus and its payload) and the means to launch it into space. The latter is what is required on the ground to design, build, and test space equipment prior to its launch into space, and subsequently to operate it in space. More often an important additional component is all the technology required on the ground to exploit what a satellite payload generates, usually called satellite data from which required information is derived to fulfill mission objectives.

An important feature of space technology is that it first needs to be qualified for functioning in space; i.e., to verify it can indeed withstand the rigours of the launch and mitigate the effects of space's harsh environment. To qualify technology for space is a long and expensive process. The usual approach is to employ those designs, materials, devices and components that have space heritage, meaning they already have a history of successful use in space. Even with proven prior space use it is a common practice to subject the equipment to rigorous and repeated testing on the ground prior to its launch.

Space equipment is composed of complex systems and subsystems built from thousands of individual components or units. It is usual to build models of these and test them before building the version that will get used for launch and operations in space. Testing is done in a systematic and prescribed manner and even a full satellite with appendages is almost always subjected to simulated conditions of launch and space prior to its launch into space.

This means that commonly available computers and electronics need to be hardened to withstand space radiation and temperatures as well as the launch conditions. As such due to the long time it takes to conceive, design, develop and build hardware and software for space, the technology being launched into space at a given time is rarely the latest that is available.

SATELLITE BUS

There are certain technologies and features that are common to all spacecraft or satellites, because certain of these functions are essential, and these elements are put together in what is called a satellite bus. Mission-specific functions are usually performed by a satellite's payload. For any satellite to work in space it is essential for it to have power generation and storage capability to ensure that onboard equipment have the means to function.

Any satellite must also have the means to always know and control its location and perhaps its orientation in space and often to change its velocity as well. Moreover, a satellite generally relies upon commands from Earth and thus has must have the means to communicate with the ground control and, for some missions, possibly be able to communicate with other satellites as well.

To maintain its required thermal profile a satellite needs the means to know and control its temperature at various locations inside and outside its body. It is imperative that a satellite is capable of managing and processing its operating status and health, and can execute commands in a timely fashion, and as well perform all the functions of a bus and payload required for a mission.

A satellite is usually composed of major systems or subsystems such as structure, power, thermal, propulsion, attitude control, navigation, and command and data handling and they together make up a satellite bus.

• Structure

A satellite's body typically has a rigid, metallic frame, that is light so as to minimize the mass that needs to be propelled up into space but also be strong enough to withstand the rigours of a launch and the radiation and thermal variations experienced in space.

Satellite structures, then, usually employ aluminum and composite materials because of their thermal control performance, structural properties and relative low manufacturing costs. The skeletal frame is often a chassis shaped as a cube, sphere, or cylinder with enough internal volume for the required equipment to be attached to it.

• Thermal Control

Most spacecraft components have a range of allowable temperatures that must be met for survival and optimum functioning. These temperatures are thus well regulated by a variety of thermal management techniques: passive or active.

Usually, passive means are employed like paints, MLI (Multi-layer insulation) covers or blankets, Sun shields, louvres, and thermal straps with thermal isolation of joints.

As well, the placement of various components and equipment within a satellite is carefully chosen so as generate the required thermal profile for each unit or device (e.g., away from the Sun-facing side for those components that are more heat sensitive). Although more expensive, many satellites also employ active means such as heaters and cryocoolers to protect against cold and heat.

• Power Generation and Storage

Satellites typically use radiation from the Sun, mainly Sunlight to generate electricity and batteries to store it. Spacecraft for interplanetary voyages employ nuclear fuels for generating electricity so as not to rely upon solar power at distances far away from the Sun.

The falling Sunlight is converted into electricity in space through the use of semiconductor photovoltaic cells, just like here on Earth, such as on the roof of a house. As each solar cell produces only a small amount of electric energy, thousands of these are mounted on panels attached to a satellite. These panels, called solar arrays are usually what is seen as long wings extending from sides a satellite.

Typically, solar arrays rotate around one axis to track the Sun so as to maximize solar radiance falling on the cells. The arrays are folded during a launch to be able to fit in a rocket's fairing and they are deployed or unfurled as soon as a satellite gets detached from a rocket's last stage in space.

There are periods when a satellite is in eclipse, i.e., when there is no Sunlight falling on its solar arrays to generate electricity. For such periods in space, electrical energy is stored onboard satellites in rechargeable batteries through reversible chemical reactions just like what we use in phones or cars.

• Propulsion

Spacecraft travelling in space generally encounter very small external forces, essentially only the gravity pulls of surrounding celestial bodies and any atmospheric drag and perhaps minute pressure from the solar radiation. However, they are normally required to alter their course, and this can be only done by applying some force. As such, they employ internal propulsion systems that provide the required force in the form of precise, short duration impulses to accelerate.

Typically, chemical propulsion through the use of liquid mono- or bi-propellant thrusters are employed, however electric and ion thrusters are used as well. Propellants are carried on board in a tank and upon a command the propellant is expelled out through a thruster that causes a delta-V (change in speed) in the opposite direction.

Thrusters are usually mounted on the aft side (back face) to increase speed. it is possible to reduce speed by turning a satellite around 180 degrees or to also mount thrusters on the fore side (front face).

- ## Attitude Control

It is important for a satellite to have a stabilized attitude and for some to even have it pointed precisely in a certain direction. It is necessary to determine and control the exact orientation of a satellite such as with respect to its flight path and/or with respect to an external entity such as the Earth or another planet.

A variety of sensors (e.g., for viewing the location of the Sun, the Earth, or a particular star configuration) are used to repeatedly measure a satellite's orientation. If this measurement is any different from what is required, then the orientation is altered to the desired attitude by applying a torque from a variety of actuators (e.g., momentum reaction wheels and magnetic torque rods) carried on board.

This dynamic adjustment is called a closed loop process whereby the orientation is continually measured, its deviation from the desired norm is determined and the remedial torque is applied. The scientific basis for dynamically changing a satellite's orientation through internal means is that the angular momentum of a body is always preserved, thus momentum is traded between electronically powered high-speed reaction wheels and a satellite's body.

Some satellites employ spin-stabilization, the simplest technique of applying spin to the outer body to maintain a steady flight path with a primary axis of orientation. However, when a precise orientation is required, spin is not enough and a more active approach is called for. As such, more satellites are actively 3-axis stabilized using a reaction wheel in each of the three dimensions. Such an approach can maintain an angular attitude to within thousandths of a degree.

• Navigation

There are typically three parts for the navigation of a satellite or spacecraft; defining a reference trajectory, keeping track of actual position while in flight, and executing manoeuvres to bring it back to a desired trajectory. A spacecraft, such as on an interplanetary voyage, can also be viewed as a satellite since its trajectory is also an orbit around some celestial body. For any Earth-orbiting satellite, once deployed by its launcher, it is in free fall towards the Earth and thereby having to go around and around at the deployed speed.

A satellite's altitude and speed are periodically measured from the ground or through GPS signals. This tracking and ranging information are typically fed into onboard and on-ground orbit propagator algorithms, a digital model that determines or reconstitutes the flight path taken and predicts the orbit to follow. Any deviation in orbit such as due to atmospheric drag is corrected through the propulsion system.

• Command and data handling

A system to process commands and handle data is essentially the brain of a satellite as it controls what a spacecraft does and when and how it is done. The commands are typically received from the ground and are distributed to the required subsystems for execution. A satellite employs radiation hardened central processing and memory units and other electronics to execute time tagged commands and also maintain health and safety. Typically, fault tolerant methodologies are employed to process transactions and manage onboard-storage.

Communications with the ground control is through onboard antennas, uplink to receive commands and downlink to transmit telemetry consisting of health and status indicators and acquired payload data to the ground for processing. Typically, different microwave frequency links are employed for telemetry and commanding and for payload data.

SATELLITE PAYLOAD - Sensors

Earth-orbiting and other satellites carry a payload to perform the functions required to realize the benefit from a space mission. A payload is typically composed of one or more sensors or instruments to make the required measurements of an object from a distance, or remotely, and thus they are called remote sensors.

These sensors can be passive or active; a passive instrument, as most are, measures naturally-occurring electromagnetic radiation such as the solar radiation reflected by an object or the radiation emitted by an object itself. While practically all space missions employ radiation measuring sensors, there are some specialized missions that employ sensors to measure gravity or magnetic fields around Earth or other celestial bodies. Active sensors supply their own radiation and measure the radiation that gets reflected or scattered back from an object.

Sensor Resolutions

All electromagnetic sensors measure radiances and get defined by what is called a sensor's resolution. A resolution is a measure of a sensor's sensitivity and capability to discern minute separations or differences in what it observes. There are four forms of resolution: spatial (separation in physical space), radiometric (separation in amplitude or brightness of radiance), spectral (separation in electromagnetic frequency or wavelength) and temporal (separation in time). In all of these, a smaller resolution is preferred because it provides finer discrimination.

Spatial resolution refers to the smallest object that can be discerned by a sensor. It is a measure of sensor's sensitivity as to how best it can differentiate, separate out or resolve neighbouring features in physical space. Smaller or finer spatial resolution is better; it shows more detail by breaking a scene into smaller units. In remote sensors, spatial resolution can be called horizontal cell size or the smallest geographical area on the ground that is imaged to compose the whole image. It is usually given by the instantaneous field view of a sensor on the ground, the smallest area from which originates the upwelling radiation that is being measured.

In digital displays there is the concept of a pixel, the smallest picture element that is displayed. Likewise, in a digital camera a scene is encoded in pixels to build the image. A higher number of pixels per unit area means a finer resolution and more details in the resultant image. For a space sensor, spatial resolution is a way of specifying its pixel size.

Radiometric resolution is a measure of a sensor's ability to discern or resolve intensity or radiance levels. The finer the radiometric resolution the more sensitive it is to detecting small differences in reflected or emitted energy. In digital displays, including digital cameras, a radiometric resolution gets defined by number of bits associated with each pixel, more bits mean more intensity levels can be displayed. Likewise, in space sensors, a radiometric resolution specifies the intensity or brightness response associated with a pixel.

Spectral resolution is a measure of how well a sensor can discern or resolve frequency or wavelength components in the electromagnetic spectrum that it receives or observes. Finer spectral resolution means narrower intervals of frequency or wavelength that can be differentiated. Thus, a sensor can breakdown the broad radiation spectrum it observes into a greater number of frequency or wavelength components.

Temporal resolution defines the revisit capability or the amount of time that it takes for a sensor to observe the same location again. Shorter revisit time intervals means the ability to discern changes that are occurring rapidly in a dynamic phenomenon. Revisit capability is typically determined by a sensor's across-track coverage on the ground known as swath width. It is the view or area that a sensor captures as it moves forward in its orbit.

Larger swath widths mean more area is imaged at a given time and the whole Earth can be imaged more rapidly again and again. A satellite in Low Earth Orbit typically makes about fifteen orbits in one day. This means to cover the whole equator in one day the swath width needs to be at least around 2670 km (Earth's circumference of 40074 km divided by 15). Typically, Low Earth Orbit imaging sensors provide revisit capability of 3 days or more especially over latitudes away from the equator. To provide a shorter revisit, missions must employ more than one satellite as is done in a satellite constellation.

The four types of resolutions are not independent and thus are optimized together. It is usual that trade-offs have to be made among these four sensitivity measures during sensor design and development, based on mission requirements. For example, measures to achieve finer spatial resolution for a sensor can result in coarser radiometric and temporal resolutions.

Passive Sensors

Passive Sensors detect and measure radiation or energy that is available from natural sources. As such, electromagnetic radiation from the Sun in the visible part of the spectrum (i.e., light that our eyes see) is most often employed, but other parts of the spectrum such as Infrared (Near Infrared and Thermal Infrared) and Ultraviolet are also used based on the phenomena to be sensed.

A passive sensor is thus a device that collects and detects naturally available radiant energy in its field of view. In so doing it measures some property or characteristic of the distant environment, target or object in its view.

All objects with a temperature above zero K emit energy in the form of electromagnetic radiation, called blackbody radiation. A blackbody is a theoretical concept of a body that absorbs all radiation falling on it, reflecting or transmitting none, and to stay in thermal equilibrium it must emit energy at the same rate as it absorbs.

As such a blackbody is a good efficient radiator and ideal emitter of energy with an emissivity of 1. The spectral distribution (the intensity of radiation over a range of electromagnetic frequencies or wavelengths) of the thermal energy radiated by a blackbody depends only on its temperature; with increasing temperatures the overall radiated energy increases and the peak wavelength decreases.

The Sun is considered to be an approximate blackbody with an effective temperature of about 5800 K and it has an emission spectrum that peaks in the central, yellow-green (590-495 nm wavelength) part of the visible (V) light spectrum (700-400nm). On either side of visible, the Sun's spectrum stretches out into infrared IR (700nm-1mm) and ultraviolet UV (<400nm) regions. The Infrared region is usually divided into three sub-regions: Near (780-2500 nm), Middle (2500-5000 nm) and Far (50-1000 μm).

As the multi-spectral radiation emitted from the Sun reaches Earth, some is reflected back to space, some is absorbed by the atmosphere, some is absorbed by Earth's surface and the remaining is reflected or scattered back upwards from its surface and is thus available to a passive sensor in space to be collected.

Of the radiation (100nm-1mm wavelength) that reaches Earth's surface, infrared radiation (wavelength >700 nm) makes up 49.4%, while visible light (wavelength 700-400 nm) accounts for 42.3% and ultraviolet radiation (wavelength <400nm) is just over 8%. Accordingly, space-based passive sensors in Earth's orbit mostly measure visible and IR radiation from the Sun coming through the Earth's atmosphere or that part which is reflected from the Earth's surface.

As a blackbody, the Earth emits its own radiation. Earth's effective blackbody temperature is usually calculated to be 252 K while the average temperature at its surface is about 288 K. The difference between the two is attributed to the presence of Earth's atmosphere and the greenhouse effect. The strength of the energy emitted depends on both the temperature of the surface and how efficiently it can emit radiation, i.e., its emissivity.

The emissivity of natural Earth surfaces is approximately between 0.6 and 1, with less than 0.85 typically for desert and semi-arid areas and above 0.95 for vegetation, water and ice in the thermal infrared range. Thus, Earth's emission is typically in thermal infrared range (around 800-1500 nm) and some also in the longer microwave (1mm-1m) part of the radiation spectrum: both are used by passive remote sensors in Earth's orbit to observe and discriminate Earth's surficial features based on differences in emissivity.

Equally, some atoms and molecules in the atmosphere have absorption and emission radiation spectra, meaning they absorb electromagnetic radiation at certain frequencies and emit radiation at other frequencies. For example, water vapour in the atmosphere has absorption bands centred at 71, 6.3, 2.7, 1.87 and 1.38 μm while those of carbon dioxide (a greenhouse gas) are centred around 15, 4.3, 2.7 and 2 μm. This physical phenomenon of an atmospheric constituent having a spectral signature is a means of detecting its presence and determining its concentration through an appropriate sensor measurement.

Examples of passive sensors are multi-spectral imagers, microwave radiometers, and optical spectrometers (or spectroscopes). These instruments employ detectors tuned to a particular set of wavelength bands typically from microwaves to infrared, visible, and ultraviolet parts of the spectrum to observe Earth and its atmosphere. Similar passive instruments based on detecting the spectral signature of constituents are also used to study other celestial bodies in the same manner.

A space-based sensor uses the motion of a satellite to view and make measurements along its ground track and undertakes some form of scanning to view and make measurements in the perpendicular or across-track direction. This results in a continual sweep of an area covering a satellite's ground track and a record of measurements over this two-dimensional swath width.

There are three approaches typically employed to scan a swath: cross-track, push-broom, and conical scanning. In cross-track a sensor has a pencil-beam shaped ground-print looking straight down. The beam is made to move side to side mechanically (rotating antenna for microwaves or a mirror for VIR sensors) to make a scan, and this scanning is repeated continually. During one scan, the sensor has moved forward and so the next scan covers adjacent area on the ground. As such, a two-dimensional measurement of a swath is produced scan by scan.

In this approach, the ground footprint of the beam at or near nadir is smaller and thus the resulting resolution is better than near the edges of the swath. The conical scanning avoids this artifact of degrading spatial resolution. In a conical scan, an instrument mechanically sweeps out consecutive arcs perpendicular to the orbital track at a fixed beam angle pointing forward. This approach keeps the ground resolution constant across the scan.

In push-broom scanning there is no mechanical motion as it is done electronically. A push-broom senor has a wide view in across-track to provide the required swath width and a narrow view in the along-track direction. The radiation from the entire view is made to fall on a fixed linear array of individual detectors, and the array is read to provide a scan and then the next scan. This line by line recording is like pushing a wide broom as a sensor moves forward in its ground track orbit.

All three scanning mechanisms are employed in space-based instruments. The scan mechanism and geometry used by an instrument affects its ground coverage such as its spatial resolution, swath width, and revisit time.

• Visible and Infrared (VIR) Imager

Visible and Infrared imagers are radiometers (measuring radiance or radiant energy) employing an array of multi-spectral detectors and they are widely used to produce detailed digital images of Earth's surface. As well, thermal infrared imagers are particularly useful in capturing thermal or temperature differential or emissivity variations of an object or a surface.

These multi-spectral imagers look down with a field of view that typically spans both sides of a satellite's ground track to cover the widest swath width that is possible. The two-dimensional image of the terrain below is produced by scanning in the across-track direction and by the advancement of the satellite in the along-track direction.

At one time scanning was done mechanically, using a rotating mirror, but it is more common now to employ electronic means by placing a linear array of detectors at the focal plane of a lens system that captures radiance over the whole swath in one go. Each individual detector measures the energy for a single ground resolution cell. A separate linear array is needed to measure radiance from each spectral band. The received spectrum is decomposed into its wavelength components with the use of a dispersive element like a diffraction grating or more typically a prism.

Thus, for each scan line, the radiance detected by each detector of each linear array is sampled electronically and digitally recorded. The movement of the satellite in the along-track direction allows radiance from adjoining ground resolution cell areas to be captured thereby recording a continuous image strip in two dimensions over the swath. Semi-conductors with materials such as silicone and germanium are used to build linear detector arrays, and some can contain more than 10,000 elements.

It was the USA (NASA) that launched essentially the first multispectral imager satellite in 1972, then called ERTS and now called the LANDSAT series. Earlier LANDSAT satellites provided digital images in 5 VIR (visible and infrared) bands at a resolution of about 30 m and thus beginning a long record of observing the Earth in such a detailed manner.

The latest is Landsat 8 (Landsat 7 continues to work as well) which operates in a Sun-synchronous 700 km altitude orbit with a repeat of 16 days. It carries two sensors and produces 740 scenes per day, each image covering an area 185 km x 185 km. One is the Operational Land Instrument (OLI) with nine spectral bands and the other is the Thermal Infrared (TIR) that has two bands.

OLI is a push-broom sensor that uses long photosensitive detector arrays, with over 7000 detectors per spectral band, aligned across its focal plane to view across the swath. The spectral differentiation of the incoming radiation from Earth is achieved by interference filters. The TIR detectors are made of gallium arsenide semiconductor chips in which electrons are elevated to a higher energy state by thermal infrared light of certain wavelengths. The elevated electrons create an electric signal that is read out and recorded to create a digital image.

In 1986, France (CNES) launched an optical imager satellite called SPOT 1 with a resolution of 10 m. Its success resulted in the SPOT series of satellites and the latest is SPOT 7 which provides multi-spectral (6 m resolution) and panchromatic (1.5 m resolution) images, each scene covering an area 60 km x 60 km. Working together with SPOT satellites in a constellation, now there are two French-Italian satellites named Pleiades HR 1A and 1B. The Pleiades satellite system provides multi-spectral image acquisition anywhere within an 800 km wide ground strip and stereo-imaging options with a 2 m spatial resolution.

The advancement of technology with the trend to provide finer or better spatial and radiometry resolutions and more rapid revisit continues as ever. Some commercial vendors are even providing image resolutions less than 1 m from their own space assets. Even the number of spectral channels available continue to increase, including hyperspectral sensors designed to measure over hundreds of narrow wavelength bands. Multi-spectral imagers working from space are now quite commonly built and operated by many organizations and many countries around the world.

• Infrared Spectrometer

In 2002, an infrared sounder instrument was launched aboard NASA's Aqua satellite to provide critical measurements in support of climate research and weather forecasting. This Atmospheric Infrared Sounder (AIRS) measures infrared brightness from Earth's surface and atmosphere. Its scan mirror rotates around an axis along the line of flight and directs infrared energy into the instrument. The sweep of the mirror creates a scan that extends 800 km on either side of the ground track.

Within the AIRS, a high-resolution spectrometer separates incoming infrared radiation into its wavelength components. Each wavelength component is sensitive to temperature and water vapour over a range of heights in the atmosphere from Earth's surface up to the Stratosphere. Multiple Infrared detectors, numbering 2,378, allow the measurement of temperature and water vapour profiles and concentrations of gases such as O_3-ozone and greenhouse gases (CO-carbon mono-oxide, CO_2-carbon dioxide, CH_4-methane).

A Fourier Transform infrared spectrometer allows measurements over a broad range of wavelengths in just one observation as opposed to the spectrometers such as the AIRS that use dispersive elements (such as a prism) to decompose the incoming radiation. A common configuration for producing a Fourier Transform is a Michelson Interferometer that generates interferograms by splitting the incoming beam of light so that one beam strikes a fixed mirror and the other a movable mirror and then brings the two beams together to produce an interference pattern from the combination or the vector sum of the two beams.

An interference pattern typically gets generated from the superposition of two waves just like the visible pattern we commonly see when we throw stones on a water surface and they generate concentric waves. When two or more concentric waves intersect, they interfere and produce regions of constructive (if a crest meets a crest the result is the sum of the individual amplitudes) and destructive (if a crest meets a trough the result is a difference of the individual amplitudes) interference in their combined amplitudes.

Within the interferometer, the movement of the mirror to an adjoining location changes the difference in the pathlength of the two beams. As a result, each wavelength of radiation in the beam is periodically blocked and transmitted due to wave interference. As such a different spectrum is produced with each displacement of the mirror. The interferograms are converted into absorption spectra through a common mathematical algorithm called Fourier Transform. This transform is a numerical means to determine the frequency components of any temporal waveform.

Interferometry (merging two or more sources of light or waves to create an interference pattern) is a common approach widely applied to make very small measurements that are not achievable by any other means. For example, gravity waves are being detected through LIGO (laser interferometry gravitational wave observatory) that can detect a change of less than a ten-thousandth width of a proton in the 4 km long path length.

Employing a Michelson Interferometer, Canada in 2003 launched a Fourier Transform Spectrometer (FTS) on board its SciSat 1 satellite as a part of Atmospheric Chemistry Experiment instrument. This mission to study stratospheric ozone chemistry in the Arctic continues to operate to date. The FTS optical instrument operates in the infrared spectral range from 2.2-13.3 μm wavelength and it uses two photovoltaic detectors cooled to 80-100 K by a passive radiator pointing towards deep space.

The SciSat 1 spectrometer looks at the Sun through the atmosphere from its high inclination (74 degrees), circular 650 km altitude orbit, in a limb viewing or occultation geometry. The instrument makes measurements twice in one orbit, during a Sunset and a Sunrise, and at different tangent heights (between 5 and 150 km). As such it produces occultation interferograms about thirty times in a day.

Each interferogram is recorded on board SciSat 1 and subsequently on the ground it is Fourier transformed into 40-60 spectra. These spectra are further processed to generate vertical profiles of atmospheric constituents (such as ozone, trace gases, about thirty organic molecules) at a vertical resolution of three to four kilometres.

- Microwave Radiometer

Microwave radiometers measure inherent radio wave radiation, emitted or absorbed, at a microwave band (mm to cm wavelengths) and are commonly used for atmospheric sounding such as for deducing temperature, gas constituent concentrations, and water vapour profiles, i.e., the variations with altitude.

The amount of radiation that a microwave radiometer receives is expressed as the equivalent blackbody temperature and is called brightness temperature. Atmospheric gases exhibit radiation absorption features and from their measurements it is possible to derive information about their concentrations around the globe at an altitude and vertical variations.

Microwave radiometers use antennas to receive electromagnetic radiation, and this radiation at such wavelengths is very weak. Thus, the signal received by the antenna is considerably amplified and the receiver is calibrated to obtain a linear relationship with signal measurement and brightness temperature. Space-based radiometers can be a sounding instrument that operates at a number of atmospheric absorption bands to determine properties of the atmosphere or an imaging instrument that produces images to characterize Earth's surface based on differences in emissivity.

For example, multi-channel radiometric measurements around 60 GHz frequency, an absorption band of oxygen, are used to derive atmospheric temperature profiles. This is possible because the emission at any altitude is proportional to the temperature and density of oxygen, and given that oxygen is uniformly distributed around the globe, the brightness temperature signals can be used to derive the temperature profile.

A 4-channel radiometer operating between 50 and 60 GHz called a Microwave Sounding Unit (MSU) was first flown in 1978 on board a satellite named Tiros-N. Nine such units as well as two advanced versions since have been used to measure temperature of Troposphere and lower Stratosphere. A similar approach is used to derive vertical profiles of water vapour, utilizing absorption bands of water vapour at 89 -190 GHz through a 5-channel Microwave Humidity Sounder (MSU).

Imaging microwave radiometers produce two-dimensional images by scanning the Earth's surface with antennas in the cross-track and the motion of the sensor in the along-track direction. An example is the Scanning Multi-Channel Microwave Radiometer first launched in 1978 on board a satellite named Nimbus.

Further since then, such instruments have in common use like a 7-channel, 4-frequency Special Sensor Microwave/Imager (SSMI) and the Moderate Resolution Imaging Spectroradiometer (MODIS) which makes measurements in thirty-six spectral bands (from 0.4 to 14.4 μm wavelengths) at varying spatial resolutions over a 2,330 km wide swath.

Recently, the Advanced Microwave Scanning Radiometer (AMSR), a 6-channel instrument with a swath width of 1,450 km has been in use for monitoring a multitude of physical phenomena. These include cloud water, sea surface temp, near surface wind speed over water, snow cover and sea ice parameters.

The Microwave Limb Sounder (MLS) measures thermal emission from the limb (edge) of Earth's upper atmosphere in six broad spectral regions to derive vertical profiles of water and ozone and several other chemical species useful for Climate-Change research. The MLS antenna from an altitude of 705 km looks forward in the orbital plane and it scans from Earth's surface to around 100 km in altitude every 25 sec.

Active Sensors

Active sensors generate and radiate their own energy or field and measure the returns upon reflection from a surface, an object or a phenomenon. Typically, these sensors employ electromagnetic energy in microwave or radio wave part of the spectrum to minimize interference from Earth's atmosphere.

Radio waves are generated when electric charges are accelerated, such as when an alternating current is applied to a metal wire, the current flow creates an electric field around the wire, and the field radiates waves outward from the wire. The wire is called an antenna, and the simplest antenna used most often is a dipole- like the common "rabbit ear" television antenna used not that long ago. Antennas are used for transmitting and for receiving radio waves.

The length of an antenna determines its radiation frequency or wavelength; shorter wavelengths require smaller antennas. Dipole antennas are used to feed parabolic or dish antennas that are seen quite commonly as they provide a more directional radiation pattern.

Some satellites also use rectangular panels on which are mounted micro-radiative elements. Together these active elements produce a directional antenna beam that has a two-dimensional pattern covering the across-track and along-track directions of a satellite's flight path. Such an antenna beam can also be electronically steered in these two directions.

Active sensors provide their own illumination from a man-made source of electromagnetic radiation, and detect and measure this radiation returned or reflected back from an object or its feature. Most common and widely used is called radar, a term coined during the Second World War for radio detection and ranging when radar sensors became first operational.

Radars are microwave instruments that use radio waves to determine the range, angle, or velocity of objects and are used to detect aircraft, ships, spacecraft, motor vehicles, weather formations and terrain. Radars are particularly useful sensors as they operate day and night, since they provide their own illumination. Importantly, they observe through practically all-weather conditions as they utilize microwave radiation that is least affected by clouds, fog, rain or snow just like the radio transmission used in any terrestrial communications system.

There are two types of radar sensors commonly employed from space in low Earth orbits: imaging (such as Synthetic Aperture Radar-SAR) or non-imaging (such as Radar Altimeter and Scatterometer).

• Radar Altimeter

A radar altimeter measures how far it is from an object or a surface such as terrain or sea surface. A satellite radar altimeter measures the ocean surface topography (surface height or sea level) by sending pulses of microwave radiation down along its ground track and measuring the time it takes for these pulses to make a round-trip from the satellite to the sea surface and back. This measurement of ocean surface height by altimeters can also be used to infer presence of large mountains on the ocean floor and to develop bathymetry (water depth) models of oceans.

A satellite radar altimeter was first employed in 1978 aboard SeaSat and since then such a sensor is practically in continuous use. The TOPEX/Poseidon mission launched in 1992 operated until 2006 and it used a dual- frequency (13.6 and 5.3 GHz) altimeter to measure sea surface height with an accuracy of 4.3 cm. Advanced versions of this altimeter were flown on Jason 1 launched in 2001, on Jason 2 launched in 2008 and the still operational Jason 3 launched in 2016.

Like its predecessor the Jason 3 altimeter operates in a 1,336 km orbit with an inclination of 66.04 degrees and it measures surface height globally every 10 days. Each height measurement is from a circular footprint about 3-5km in diameter along the nadir (looking down) ground track. This measurement is useful for many oceanography applications such as observations of ocean waves, tides, currents, sea ice and glaciers over land.

• Radar Scatterometer

Polar orbiting scatterometers are microwave radar sensors used to measure the reflection or backscatter of microwave energy while they scan Earth's surface. The microwave backscatter is particularly influenced by surface roughness on the scale of a wavelength, i.e., in the order of centimetres. This property is used to estimate wind speed and direction from the backscatter measurements made by a satellite-based scatterometer. Winds produce surface roughness over water and the microwave backscatter is seen to increase with increasing wind speeds due to increasing surface roughness.

NASA's SeaSat in 1978 was the first satellite to demonstrate the value of a scatterometer and since then many such instruments have been flown. Both of ESA's ERS 1 (launched in 1990) & 2 missions had the Active Microwave Instrument (AMI) sensor which operated in a scatterometer mode.

NASA's QuikSCAT mission launched in 1999 operated for more than ten years and it had on board the SeaWinds scatterometer operating at a frequency of 13.4 GHz in an 802 km polar orbit. This scatterometer employed dual-beam conical scanning at fixed 46- and 54-degree angles that provided an 1,800 km wide swath on both sides of the nadir and a daily global coverage. The scatterometer had a 25 km ground resolution cell and the measured backscatter provided an estimate of wind speed and direction at 10 m above sea surface.

The Advanced Wind Scatterometer (ACAT5) onboard Eumetsat MetOp satellites (starting in 2006) operates at a frequency of 5.255 GHz. It improves upon ERS 1 and 2 by providing a daily global coverage at a nominal resolution cell size of 50 km. It has 550 km wide double swath, one on either side of satellite's ground track. The swath is produced through three antennas on each side in a fan beam configuration. This scatterometer produces continuous backscatter measurement for the estimation of wind speed and direction, sea ice and snow cover, and soil moisture.

The Precipitation Radar aboard Tropical Rainfall Measuring Mission (TRMM) was the first satellite instrument to provide a 3-dimensional map of storm structure. Launched in 1997, the TRMM operated for seventeen years and provided a daily global coverage from a nominal 175 km altitude, 35-degree inclination orbit.

Employing microwave transmission at a frequency of 13.3 GHz and measuring radar backscatter, the Precipitation Radar provided 3-dimensional rainfall distributions over land and ocean surface. It had a horizontal resolution of around 5 km and a 247 km swath width. It also provided vertical profiles of rain and snow from the surface up to a height of about 20 km with a vertical resolution size of about 250 m.

- Radar Imager

Synthetic Aperture Radar (SAR) Imagers are so called because they employ satellite's motion to create a synthetic long antenna (several kilometres in length) in the along-track direction, much longer than the physical antenna being carried on board the satellite. Larger antennas produce finer or better spatial resolutions, meaning they provide more detailed images, being able to highlight, detect, and identify smaller and smaller features or objects.

Given that the size of a physical antenna that can be carried aboard a satellite is limited, synthetic aperture technique is of immense value in producing fine spatial resolution images and resolving fine features of terrain below from satellite altitudes. SARs typically look to a side of the satellite ground track and its physical antenna provides a wide beam in the cross-track direction to constitute the swath width.

A SAR transmits a number of microwave radiation pulses (shorter pulses provide finer cross-track spatial resolution), each pulse hits the ground and gets scattered back, the pulse return from the nearest range of the swath arrives back first in time and the one from the farthest range arrives last. Each pulse return is thus a scan of the ground swath and each successive pulse transmit and receive cycle produces successive scans while a satellite is moving forward in its orbit.

The time sequence of pulse returns results in the production of a two-dimensional radar image showing intensity of the signal backscattered from the terrain below. A SAR is a coherent instrument, meaning it needs to maintain an alignment in phase between the transmit and receive pulses and must record both the phase and amplitude of the return signal. This phase history allows a large synthetic antenna to be created later on the ground through combining successive pule returns using digital processing algorithms and producing a high-resolution image of the terrain overflown.

Because SARs preserve the phase history, SAR image data are unique for deducing topographic changes of the order of a wavelength or less through coherence change detection from multi-pass observations. This is done through InSAR (interferometric SAR) technique where two SAR images of the same area acquired at different times are used to produce an interferogram, through differencing them, which displays the ground-surface displacement (range change) between the two time periods.

This uniqueness of SAR observations for deducing minute displacements in terrain features is of immense utility for a variety of applications. Space-based SAR measurements are being applied to practical uses such as monitoring land subsidence operationally over oil and gas well sites and even deducing the slow movement of a glacier that is just a few centimetres to metres per year.

The first space-based synthetic aperture radar imager was on board SeaSat launched in 1978. This satellite operated only for about 100 days, but by providing all-weather, day/night observation capability at a frequency of 1.275 GHz with a spatial resolution of about 25 m it proved the value of such an instrument in space.

ESA then launched in 1992 a satellite called ERS 1, followed by ERS 2 (1994) and ENVISAT (1998) and each had on board a synthetic aperture radar imager. These SARs operated at a frequency of 5.3 GHz and provided imagery at a spatial resolution of 30 m over a 100 km swath. ESA has achieved SAR data continuity as well as improvements in SAR performance through the launch and operation of the multi-sensor Sentinel series of satellites starting in 2014.

Canada launched in 1995 the first multi-mode synthetic aperture radar, called RADARSAT, that also operated at a frequency of 5.3 GHz and provided images at a variety of resolutions (5 to 25 m) and swath widths (100 to 500 km). Its antenna in space measured 15 m in along-track and 1.5 m in cross-track and was built in four panels that were folded during launch. RADARSAT 1 provided global coverage operating at a nominal altitude of 800 km in a sun-synchronous polar orbit with an inclination of about 98.6 degrees.

The multi-mode capability of RADARSAT 1 allowed a user to select parameters of SAR images, such as the incidence angle, spatial resolution and swath width, which would best suit the intended geophysical application. This versatility of application specific observations was found to be extremely useful by users and it resulted in many operational applications around the world.

Earth observations satellites that carry passive optical instruments, typically cross the equator at mid-morning or early afternoon local time. This is to ensure that there is enough sunlight and that cloud cover is minimum. However, active sensors such as SARs do not have such constraints. As an innovation, RADARSAT 1 crossed the equator at dawn-dusk (6 a.m. or 6 p.m.). This allowed it to reduce demand on battery power as its solar arrays could be in sunlight during almost the whole orbit for most of the year.

RADARSAT 1 was also the first to use a technique called SCASAR through which a SAR provides a much broader swath width (up to 500 km wide images) than normally possible. This was a considerable improvement over earlier SARs that essentially had a narrow swath of the order of 100 km. Also, the orbit had an exact repeat of 24 days but had a 7-day primary sub-cycle (the interval between the passes over adjacent ground tracks) and a 3-day secondary subcycle. RADARSAT 1 thus provided daily coverages of the Arctic (a prime mission requirement) and a 3-day repeat coverage of areas near the equator.

RADARSAT 1 was the first to map in detail the whole of Antarctica, in 2000 and then again in 2003. To undertake this, the satellite had to undertake a 180-degree yaw manoeuvre, to look to the left, instead of the usual right looking orientation, so as to fill the gap over the south pole. Through this historic imaging, it became possible to estimate the small motion of advancing glaciers in the Antarctic and the loss of this ice cover and thereby deduce the early effects of Climate Change.

Though designed for a nominal five years life, RADARSAT 1 mission lasted more than seventeen years as it continued operations until 2012. However, RADARSAT 2 was launched earlier in 2007 to continue multi-modal SAR capability as well as to provide enhancements such as multi-polarization images and an improved resolution of 3 m. Further in the series, RADARSAT Constellation Mission (RCM- a constellation of three satellites) was launched in 2019 not only to continue SAR data continuity but to substantially build upon the capability of its predecessors.

From a 600 km altitude, RCM's three satellites operate in the same orbit plane phased 120 degrees apart, providing a more frequent coverage than RADASAT 1 & 2. Moreover, it has coherent polarization (polarimetric) and coherent change detection capabilities that allow finer and additional discrimination of geophysical characteristics. Each RCM satellite also carries a secondary payload of AIS (Automatic Identification System), essentially a receiver for tracking ships, given that ships are required to carry a transmit beacon that allows their detection from satellite altitudes.

Germany (DLR) launched a SAR mission operating at a frequency of 9.6 GHz on a NASA space shuttle in 1994 called SIR-C/X and again in 2000 called Shuttle Radar Topographic mission. Germany has since continued this innovation of SAR stereo coverage for producing topographic maps with its TerraSAR series of satellites (TerraSAR-X and Tandem-X in 2010 and 2014). Italy (ASI) launched its COSMO Sky-Med SAR constellation working at a frequency of 9.65 GHz starting in 2007 and India (ISRO) launched a SAR satellite in 2015.

SAR technology therefore continues as ever to advance and more nations and organizations such as South Korea and UK are launching and operating their own SAR imagers. This is a testament to the utility of SARs for Earth Observations as they offer an operational high-resolution, all-weather, day/night imaging capability that is vital for so many applications.

GROUND INFRASTRUCTURE

Ground infrastructure for a satellite system typically consists of a mission control to task and operate a satellite and facilities to receive, process, and archive payload data and provide sensor calibration. Included are vital ground elements such as networks of antennas to reliably communicate with a satellite and to receive what a payload has acquired.

The most critical ground element is a Telemetry, tracking and commanding (TT&C) facility that has redundant antennas at its disposal. A tracking TT&C antenna follows a satellite in its orbit as it first comes up and then disappears below Earth's horizon. Such an antenna is used as a link between a mission control and a satellite by employing radio waves typically around 3 GHz frequency.

Radio signals are transmitted and received to communicate with a satellite while it is being tracked by the TT&C antenna—an uplink for sending commands and a down link for receiving telemetry sent by a satellite to mission control. Uplinks are also used by mission control for uploading any newer version of the onboard software.

Telemetry provides the health status of a satellite as well as the status of the commands executed by a satellite. An important function of TT&C is to provide ranging information about a satellite's orbit; its altitude is determined from the time delay between sent and received signals and its speed from the Doppler shift (like a change in pitch of the whistle from an approaching train) in the frequency of the return signal. This range measurement is used in regularly updating the orbit or the trajectory of a satellite.

Ground antennas are also required to receive payload data sent by a satellite. Radio wave links at a high frequency (typically around 10 GHz or higher) are used for payload data transfers so as to accommodate high data rates in the range of 300-600 million bits per second that are required. Higher radio frequencies or even optical links using lasers are attractive as they allow broader bandwidths and higher data rates but these usually suffer operational drawbacks such as disruptions from atmospheric conditions.

Both the TT&C and payload data reception functions are sensitive and need to be properly secured. Thus, encryption of any link between ground and a satellite is increasingly becoming a standard practice so as avoid the possibility of hacking.

A network of ground stations exists around the world, geographically well placed and some providing commercial services, for uplinking of commands to a satellite and downlinking of telemetry and of payload data when required. These antennas are typically parabolic dishes around 10 m in diameter.

However, some TT&C antennas need to be extremely large, for example 70 m or more in diameter, when employed to communicate with a spacecraft far from Earth. They need to be sensitive enough to pick up week signals from deep space and powerful enough to send strong signals, such as to communicate with a spacecraft travelling to other planets. While it takes only about 2.5 sec for radio signals to travel from Earth to the moon and back there is a considerable delay for other planets. One-way radio propagation time from Earth to Mars ranges from 4 to 24 min and to Neptune, the farthest planet, it is more than 4 hrs.

An important part of ground infrastructure is to rapidly archive and process the payload data and extract the information from it for exploitation by users. The amount of data that gets acquired through space missions is vast, meaning all these data need to be stored for a long period of time, often years if not indefinitely, and be accessible for rapid processing and analyses.

It is the payload data that is a vital economic return from vast investments required in any space mission, and thus safeguarding this data asset for scientific, operational or commercial exploitation is of paramount interest and importance. The technology required for digital storage and processing is improving every day, reducing storage media volume and its cost while increasing speed of access and processing.

Ground infrastructure is a vital component of any space mission, and great care is undertaken that ground assets installed are responsive to mission requirements and can serve multiple missions. Thus, standards are developed and applied through inter-space agency panels to create a compatible global network, even though parts are owned and operated by different organizations.

ROBOTIC EXPLORATIONS

In order to undertake robotic explorations of the moon or a planet, a number of specialized technologies are required. Perhaps the most critical component is the means for a safe deployment of the robot from an orbiting satellite to the surface of the body to be explored. This is challenging as the descent is typically through an environment quite different from Earth such as of low gravity and rapid atmospheric or thermal variations.

As well any deployment of a robot onto a body's surface needs to be programmed and timed as an automatic activation, a pre-recorded sequence. For deployments through an atmosphere, mechanisms such as a parachute are found to gradually slow the descent. Carefully wrapping a robot with cushions can allow a safe landing.

In addition, once landing safely on a body's surface a robot needs to survive and perform its desired functions with minimal supervision from Earth. Tele-robotic operations from Earth require compact, fault tolerant onboard sensors, means for power generation and storage, need for reliable communications links, capability to function autonomously and execute near real-time commands, and capability for safe locomotion on some difficult terrain.

HUMAN EXPLORATIONS

Given that space is a harsh environment, the use of robots is the preferred means for space exploration, on account of relative cost and disposability. However, for certain missions, human explorations may enhance the scientific value. But ensuring that life is protected in space remains a most daunting challenge due to the need for a fool-proof life support system.

First, there is the need for a space suit to protect against the outside harmful radiation and to regulate the internal environment. It must supply oxygen-rich air and keep the temperature, humidity, and pressure within the appropriate range for the human body. Then there must be a means for an emergency evacuation of humans and their safe return to Earth.

The means for safeguarding humans in space has evolved significantly through earliest days. More is understood now of the impact of space on the human body from astronauts spending their time in a space habitat during the space shuttle and space station missions.

However, for long duration space excursions that will last several years there remains caution. There is still a need to address what and how to provide for one human or a group of humans the mental, physical and nutrition nourishment in a safe and reliable manner in a small, closed container far from Earth for a long period of time. Research continues to find appropriate solutions for these challenges and others like how to make space voyagers survive in an alien environment and to deal with emergencies that will undoubtedly arise.

8. SPACE APPLICATIONS AND USES

Space missions are undertaken to serve a purpose and meet a need. Typically, these missions are conducted for a utilitarian application such as for Communications, Navigation, Earth Observation, Space Science, Astronomy, Solar and Planetary Exploration, and Human Exploration. It is important to underline that while a vast majority of space applications are for civil and peaceful purposes, space is also attractive and usable for military purposes; most of the developed space technology applications can be of dual-use.

Since the launch of Sputnik 1 in 1957 by the Soviet Union, about 8900 satellites are reported to have been launched to date of which around 1900 of these are still operational serving a variety of applications or uses. It is estimated around 63% are in Low Earth Orbit typically for Earth Observation applications and 29% in Geostationary Orbit typically for communications applications around Earth. The remainders are in medium or elliptical orbits typically providing navigation or communications services.

TELECOMMUNICATIONS (Satellite Communications-SatCom)

Since the launch of the world's first communications satellite called Telstar in 1962, use of satellites to provide communications around Earth is the most common and commercial space application with about 2000 of these satellites in Earth's orbit. A satellite with a transponder or a repeater provides a unique and efficient vantage point to link any two communicators on Earth, irrespective of the distance between them and overcoming Earth's curvature.

A communications satellite receives a message in the form of a series of radio pulses from Earth, and it typically amplifies these pulses and relays forward the same message back to Earth. In order to handle a large volume of messages that can be handled simultaneously it is necessary to have a broad bandwidth radio link between Earth and a satellite. It is also essential to minimize any perturbations caused by Earth's atmosphere to radio signals as they travel to and from a satellite. Typically, frequencies around 5, 15, 30 GHz are employed to minimize size of the antenna required and to also allow broad bandwidths.

Due to their utility, practically all countries have national communications satellites for broadcasting television and radio signals, providing telephone links, and even internet to their citizens. Communications satellites use radio waves or microwaves and many channels within the same frequency band and employ packaging of signals (time and frequency wise) to service a multitude of users at the same time and these are seamlessly integrated with terrestrial telephone or television networks. The vast majority of these satellites are in geostationary orbit, while polar orbits are also being employed.

The geostationary orbit around the equator is becoming congested; the satellite slots are licensed by International Telecommunications Union (ITU). There are consortiums that employ satellites in geostationary and polar orbits to provide international communications services on land and on oceans around the world. Internet services based on satellites are the next steps in global communications.

NAVIGATION (Satellite Navigation-SatNav)

Signals from geo-positioning satellites (GPS) have revolutionized navigation on Earth. The USA was the first to launch, starting in

1978, a GPS network, first to meet military needs and subsequently for civilian use. This network is comprised of twenty-four satellites, uniformly distributed four each in six elliptical orbits—additional satellites are put in orbit to deal with contingency. These GPS satellites orbit the Earth at around 20,000 km in altitude, making one orbit in about twelve hours.

Now, there are regional GPS networks around the world such as by the European Union, Russia, China, Japan and India to augment the USA network. These satellites continuously transmit precisely timed, encoded radio frequency signals towards Earth. The two frequencies used by GPS are 1.575 GHz and 1.227 GHz so that signals can penetrate cloud, fog, rain, storm and vegetation.

Thus, GPS based location devices receive accurate signals in all-weather conditions. These devices use complex algorithms to process signals from at least four satellites at a time. The distances between the satellites and the given device are computed allowing the algorithm to determine the location or coordinates of the device in a dynamic fashion.

GPS-based navigation and location maps are now common in surface vehicles, smart phones and other devices on Earth and are even used to determine location in Earth-orbiting satellites. The accuracy of the location is typically around 1 to 2 m and is determined by the grade of the signal and processing algorithm; military grade signals provide better service.

EARTH OBSERVATION (EO)

Earth-orbiting satellites, typically in low Earth orbits are particularly suitable and effective to look towards the Earth and provide useful, timely and up-to-date information on prevailing conditions in atmosphere, on oceans, and on land. The repetitive timeliness of satellite observations is of value for deducing a physical phenomenon at any temporal scale whether its rapid for weather or moderate for environment and land resource or really slow for climate processes.

EO satellites typically employ a variety of electromagnetic remote sensors depending upon the application and phenomena to be sensed. Most sensors are passive as they measure solar radiation reflected from Earth's surface towards them. Active sensors employ their own electromagnetic radiation and measure this radiation that gets scattered back from the Earth's surface.

• Geophysical Information Extraction

In either case, passive or active sensor, the real challenge is to understand and formulate the physical process of radiation coming from Earth's surface incorporating any effects of the intervening atmosphere. Thus, in all EO applications a major enterprise is to deduce or derive geophysical parameters from the sensor's measurement of radiation.

There is a variety of geophysical information of interest dependent on a mission's objectives. This includes such information as the concentration of a gas in the atmosphere, the type and vigour of a vegetation, or the amount of moisture in the soil, all to be derived from sensor measurements. To be of value, these conversions are preferably done in as real-time as possible. This requires fast processing power given the vast amount of sensor observations that get acquired on a daily basis for most of these applications that are essentially global in scope.

A key aid in converting any sensor recording from engineering to geophysical units is the idea of change detection. The main value of satellite observations lies in being able to look at any location on Earth in a repetitive manner, from constant surveillance through sensors in geosynchronous or geostationary orbits to temporal coverage of only hours from sensors in polar orbits. This results in a temporal record that allows changes of an object or phenomenon to be deduced.

For example, when an agricultural field is irrigated, it results in an increase in soil moisture and that change is detected from space-based measurements by comparing images of the field before and after the irrigation. The capability to measure temporal changes that occur is a powerful technique of value in virtually all EO applications. For example, a temporal signature is used to detect and identify a crop type by combining three to five observations of a field over the crop calendar, from plantation to harvest, a cycle that is typically of several months in duration.

Another key value of satellite-based observations is the capability to view coincidently the same location or a phenomenon with more than one sensor or at more than one frequency band. Different sensors or frequency bands are responsive to different geophysical parameters and thus combining or fusing multi-sensor or multi-frequency or even multi-polarization observations is of tremendous aid in detecting and identifying physical processes and changes therein.

For example, radar backscatter is particularly influenced by surface roughness and moisture content of terrain as well as the subsurface penetration of radar signals which produces volume scatter; all these parameters are wavelength dependent. Thus, an agricultural field observed at different frequencies by radars highlight different features or conditions and combining or fusing these images results in a multi-spectral signature giving a more complete description of the state of its vegetation cover.

As such, application-specific algorithms are employed based on considerable research and development efforts to deduce geophysical properties from sensor measurements. These measurements can be acquired by just one sensor or in combination with other sensors, at one time or a multitude of times. In all cases, considerable care is taken to continuously calibrate sensor measurements and to validate information extraction algorithms so as to ensure that the most accurate geophysical information is obtained.

• Weather (Weather Satellites)

Satellites dedicated to weather observations in geostationary orbits became operationally early in the Space Age through the USA. For example, launched in 1960, TIROS (Television and Infrared

Observation Satellite) was not only the world's first weather satellite but it was also the first to provide satellite imagery of the Earth. Since then TIROS series has continued with forty-five satellites in operation, modified and improved over the years and supplemented by other weather satellites.

Weather observations and forecasting is a demanding application given the need to measure a variety of different physical parameters such as cloud cover, winds, precipitation, and solar radiation flux in addition to temperature and pressure variations with altitude over both land and ocean. Moreover, the physical processes are complex and dynamic and cover large areas; some phenomena change over a time frame of minutes and some can be continent-wide or even global. Such a wide scope results in the deployment of models that require vast volumes of measurements and data intensive computing.

Space-based observation sensors are unique in providing not only synoptic global view and measurements of various physical processes at local, regional, and global levels but also doing so at the required temporal resolutions such as even less than an hour to record any rapid, dynamic phenomenon. These measurements combined with employment of fast computational resources are thus vital in the provision of hourly, daily and weekly weather forecasts for any location on Earth.

As such, weather satellites using a suite of sensors from imagers (e.g., VIR and microwave) and profilers (e.g., microwave radiometers) now form the backbone of operational weather services, supplementing the terrestrial measurement network.

These sensors from polar or geostationary orbits continuously measure a variety of physical parameters such as clouds and precipitation, temperature, pressure, wind profiles with altitude, amount of Sun's energy falling on Earth and that reflected back, ocean currents and waves, ocean height, ice cover, land cover, snow and its water equivalent, and soil moisture. All these and other parameters are required as input to numerical weather models.

Weather satellites operating in geostationary or polar orbits are unique to provide timely, global coverage especially over wide swaths of open oceans; all these regions not only have sparse ground-based measurements but they drive the global weather patterns. The USA (NOAA) and Europe (EUMETSAT) have fleets of such satellites used in weather now-casting and forecasting.

Among more than twenty satellites that NOAA typically operates there are Geostationary Operational Environment Satellites (GOES) for short-range warning and now-casting and Polar Orbiting Satellites for long range weather forecasting, atmospheric and weather monitoring; each equipped with a variety of sensors. Likewise, EUMETSAT operates Meteosat and MetOp series of satellites, again each comprised of multiple sensors. Other countries such as Russia, China, Japan and India have their own satellite assets for local or regional weather watch.

Observations from weather satellites are generally shared among nations and there are organizations such as the World Meteorological Organization to facilitate and foster regional and international cooperation.

• Mapping (Satellite Images)

Digital spectral images, from an optical or a radar sensor alone or together with other space derived information such as geo-location, are becoming the foundation for producing and updating practically any map of Earth. Such observations from space are unique in providing the spatial extent and through repeated coverage the temporal extent of any phenomenon on Earth.

Further, spectral images are valuable in highlighting, detecting and even identifying particular features such as vegetation type and vigour, mineral deposits, areas for fishery or the extent of flooding. These images, taken through one- or two- pass stereo coverage, can also provide topographic information such as height or elevation with high accuracy.

Microwave sensors are particularly useful for observing those areas on Earth that are often cloud or fog covered and those geophysical phenomena that are dynamic and changing rapidly over time. This means that some applications need reliable observations or measurements of the same area or location repeatedly over time. The all-weather, day/night observations and moderate to high spatial resolutions capability of microwave instruments are particularly essential for these applications and certain regions on Earth.

For example, satellite radar images are routinely and operationally used in ice surveillance mapping such as in the Arctic and Antarctic which are marine areas that often have poor visibility and suffer long periods of darkness. With these images it is possible to map ice concentrations and floe sizes and delineate ice type and thereby its thickness and strength. This is vital for navigating in ice covered waters such as through the North-West passage in the Arctic, an important economic shipping lane.

Equally, radar imagery is operationally applied in detecting oil pollution and its extent on ocean's surface. Oil slicks dampen the wind induced capillary waves and thus the surface roughness, thereby affecting the radar backscatter and allowing the delineation between oil slick and surrounding open water. Importantly, radar imagers are operational surveillance tools for detecting illegal dumping of oil in oceans and thereby polluting fishery and marine life. Such monitoring is also valuable in identifying offshore areas of natural oil seepage that may contain sub-surface oil fields warranting a more extensive exploration.

Space derived information is now vital for undertaking and updating of inventory of renewable resources such as agriculture and forestry. For example, maps of agricultural areas delineating crop type and field conditions (e.g., soil moisture and vegetation growth) from satellite images are used for estimating crop yield and as well for precision farming to improve production.

In forestry, these space images can delineate clear-cut areas undergone logging and the success of regeneration management. In geology mapping applications, space imagery provides information on surface rock lineaments, folds, faults and topography as well as surficial vegetation of value in mineral and oil explorations.

Space-based digital spectral images are also crucial in land use mapping and urban planning. These high-resolution optical and radar images provide detailed information on land cover and topography useful for deducing transport routes and energy transmission corridors and making zoning decisions in terms of housing and commercial sectors. There are now available a variety of analytical tools such as Geographic Information Systems that combine or merge space imagery information with social data bases such as from census and are vital for investment decisions and political governance.

• Environment Monitoring (Satellite-based Surveillance)

Global monitoring of the atmosphere, oceans and land has become routine through the measurements of a variety of parameters daily by a string of instruments on satellites looking down towards the Earth. Starting in 2005 there is a coordinated international approach to develop a Global Earth Observation System of Systems (GEOSS) that links contributions of satellites, instruments or systems from various national or international organizations.

The goal is to have a comprehensive, up-to-date, and timely observation capability to look at Earth as a whole in a detailed and systematic manner and document changes effecting Earth's biodiversity, depletion of resources and quality of its environment.

As a part of GEOSS, the USA (NASA) and international partners have a Morning Constellation (for observations each morning at equator crossing times around 10:30 a.m. local solar time) and an Afternoon Constellation-dubbed A-train (for afternoon observations at equator crossing times around 1:30 p.m. local solar time)– a series of satellites that follow each other so as to measure the same feature or area with different instruments and get as coincident or simultaneous measurements as possible.

The overarching objective is to have a long record of systematic measurements documenting Climate Change and its impacts that allows evidence-based decisions to be undertaken on a global basis for mitigation and recovery efforts.

To address the challenges posed by Climate Change, ESA and EU have been undertaking a comprehensive Earth Observation program, building on earlier ERS and ENVISAT missions. This EO program is called Copernicus under which a series of Sentinel satellites is developed, launched and operated. Each satellite is fitted with a variety of sensors and makes measurements globally.

These Sentinel satellites support many diverse applications by recording changes in the Earth's surface and its environment. Most of these applications are operational in nature, meaning a lot of effort is invested to rapidly process the acquired satellite data and put the geophysical-derived information into the hands of scientists and other users on a near real-time basis.

As such, through these international efforts a global space-based observations and information system is being established to monitor Earth's cryosphere, geosphere, atmosphere, biosphere, and hydrosphere and thereby to study the Earth as one whole system. An early indicator of global warming and the impact of Climate Change is attributed to melting of sea ice and glaciers. Space-based radar observations have been vital to document depletion of sea ice in the Arctic and the advancement of glaciers in Greenland and Antarctica into surrounding oceans.

As an example, Canadian RADARSAT 1 satellite images resulted in the production of annual Arctic ice atlas starting in 1996 and this mission was first to produce a detailed map of Antarctica in 2000 and again in 2003. SAR imagery thus produced was able to measure the depletion of sea ice in the Arctic over time and as well document the slow motion of ice glaciers (typically metres per year) advancing into the Antarctic ocean.

Routine mapping of sea ice and land ice continues with RADARSAT and other missions to record the rate of change in the Earth's cryosphere. Rapid melting of frozen water on Earth due to temperature increases from Climate Change is projected to result in an accelerating rise of global sea level impacting low-lying islands and coastal areas.

Other than ice mapping, RADARSAT 1 also pioneered under its Background Mission, a systematic acquisition of cohesive global data sets of long-term scientific value for environment and Climate-Change research such as continental wide coverages and mapping all the low-lying islands repeatedly. Many other missions, such as the LANDSAT series, also have long historical records of global observations of value in studying changes to our planet Earth over the recent past.

• Disaster Management (Satellite-based Support)

The use of space derived information for protecting life and property on Earth from natural disasters such as floods, Earthquakes, landslides, forest fires, etc., is perhaps the most useful or public good space application. In 2019 alone, natural disasters are estimated to have caused an average economic loss of about 232 billion dollar and thousands of deaths around the world, and these events are becoming more frequent and more severe due to Climate Change.

Disaster Management is a cyclic activity comprised of phases: prevention, preparation, operations, and clean-up or reparation. Space derived information such as from digital spectral images is essential and useful in each phase of a disaster.

Satellite observations however are especially crucial during the operations phase, from the time when a disaster event is just starting to the time when the emergency is declared finished. During this period the contributions from satellite sensors are unique as they witness the progression in the extent and severity of a disaster's impact on Earth. This timely information allows rapid and pertinent deployment of relief operations and is vital in managing a disaster effectively and efficiently and in minimizing loss of life, property, and economy.

To be able to effectively support management of a natural disaster anywhere in the world, some satellite operators have banded together and established an operational system called International Charter Space and Major Disasters. The Charter was conceived by ESA and CNES and joined by CSA and the three founding agencies worked together to build an operational framework and then to start Charter's operation on Nov. 1, 2000.

The Charter was formulated in response to the recognition that space-based multi-sensor, multi-mission observations offer the only viable, timely means to cover the wide variety of disasters that occur around the world. Also, the fact that no single nation can have enough space assets of its own to monitor a disaster when it occurs. Therefore, cooperation among nations and agencies is vital to minimize economic impact and save lives, property and resources.

Now the Charter has seventeen member agencies from around the world and through their combined effort space images and derived information are provided to disaster relief organizations on a priority access basis and free of charge whenever there is a major natural disaster anywhere in the world. Since its operational deployment in 2000, the Charter has responded to more than 650 disasters in 126 countries such as for floods, earthquakes, landslides, hurricanes, tsunami, oil spills, and forest fires.

For each disaster event the member agencies prepare and execute a data acquisition plan to acquire and then process images from more than sixty contributing satellites and deliver these images to disaster management authorities within hours.

This is a 24/7, 365 day a year, coordinated space operation spanning the whole globe. The Charter acquires satellite images that are best to reveal a disaster's impact given that each kind of disaster, like a flood, has a combination of sensors that furnish the most information. The Charter now has a unique 20-year historical record of global disasters and their impacts, a scientific database useful for disaster preparation and prevention.

SPACE SCIENCE

There are a variety of research activities under the wide scope of Space Science. These include study of our Sun and the associated solar system, the Sun's impact on our Earth, space environment, astronomy, explorations of our moon and other planets and their moons, study of comets and asteroids, study of the impact of microgravity on humans, animals, plants and on materials, liquids and gases.

• Solar Research

As to be expected our Sun is the most observed and studied body in space due to its vast influence on Earth and on the rest of the solar system. Solar research among others is concerned with understanding the interactions of Sun's energy with Earth's atmosphere and magnetic field and such solar- terrestrial phenomena as Aurora Borealis and changes to Earth's Ionosphere.

The use of satellites to study the Sun commenced right at the beginning of the Space Age. NASA started the Orbiting Solar Observatory series in 1962 and has operated eight of these space telescopes. In addition, since then a countless number of satellites have imaged the Sun and measured its spectroscopic characteristics.

To date, there have been around 100 solar-dedicated spacecrafts plus hundreds of others with solar instruments. The USA has launched and operated the greatest number of satellites for solar research while others such as Russia, ESA, and Japan have also contributed significantly.

The Sun continues to be the focus of several current and recent missions that are observing the Sun from really close-by distances such as NASA's Parker Probe (launched in 2018) which is less than ten solar radii away and ESA's Solar Orbiter (launched in 2020) which is around sixty solar radii away. To put this in perspective, Mercury, the planet nearest to the Sun, orbits at about sixty-nine solar radii while Earth's nominal orbit is at around 215 solar radii.

• Space Environment

Solar flares or mass corona ejection from the Sun have significant effect on Earth-orbiting satellites and voyaging spacecraft. For example, these charged particles are known to disrupt telecommunications and power lines and other equipment on Earth. As such study of space weather that includes observations of solar output as well as cosmic radiation are essential to prevent harm to our space assets and terrestrial infrastructure.

The solar output seems to follow an eleven-year cycle going from minimum to maximum. NOAA operates a Space Weather Prediction Centre that issues watches, warnings, and alerts for hazardous space weather events so that precautionary measures can be undertaken.

• Astronomy and Cosmic Research

From Galileo's first ever use of a telescope in 1609 to look skyward and observe our moon, Jupiter and the Milky Way, countless telescopic observations of celestial bodies have been made since then. A number of large telescopes have been established on Earth, alone or in a network spanning the globe, to observe our universe across the full electromagnetic spectrum from gamma rays, visible and infrared, to microwaves and longer wavelength radio waves.

There is an increasing international cooperative effort to undertake a systematic study of the Cosmos; the distant stars, galaxies, blackholes, and other celestial objects and phenomena. However, it is known that Earth's atmosphere causes perturbations and thereby limits what can be observed from Earth's surface looking up. As such space telescopes operating outside of Earth's atmosphere are especially valuable in astronomical and cosmic research.

Since the launch of the first space telescope in 1965, called Proton 1, by the Soviet Union more than 100 such instruments have been launched by many countries. These telescopes range in operation from gamma rays, x-rays, ultraviolet, visible, infrared, microwave, and radio waves to cosmic rays and particles. Most of these instruments are placed in Earth orbit while there are a few on lunar surface, in Solar orbit, and at the Sun-Earth Lagrange points.

The most notable and well known is the Hubble space telescope, which was launched in 1990 and is till operational today. Hubble operates at UV and VNIR (0.1-1 μm) wavelengths and its high-resolution images of objects in deep space have popularized and revolutionized astronomy. As a successor to Hubble, the James Webb space telescope (JWST) planned to be launched in 2021 will be the largest and most powerful space telescope.

Comprised of a Sun shield measuring about 20x14 m, JWST's mirror will be 6.5 m in diameter, much larger than Hubble's 2.4 m mirror. JWST will observe in lower frequency range from long wavelength infrared to midinfrared (0.6-28.3 μm). The telescope will be placed near the Earth-Sun Lagrange point and its Sun shield will keep the mirror and instruments very cold to below 50 K temperature, allowing it to see farthest into deep space.

On the other end of size scale, the first smallest space telescope was Canadian, called MOST (Microvariability and Oscillations of Stars), launched in 2003. It operated until 2019 and It just weighed 53 kg and measured 60 x 60 x 24 cm and used a visible light dual CCD (charge couple device) camera.

These and future telescopes in space will continue to view and study our universe, its beginning - the Big Bang, the extent and expansion of space, black holes, the variety of stars, galaxies, expoplanets, dark matter and energy and other such phenomena. The aim of all is to explore if there is life elsewhere in the universe and to understand Earth's place in the Cosmos.

SPACE EXPLORATIONS - Planetary Voyages

It took nearly 50 years of human endeavour to have spacecraft visit all the planets in our solar system; almost all the spacecraft were launched by the USA. This space exploration venture began with planet Venus in 1961 and it was completed with Pluto in 2015. In

1961, the Soviet Venera 1 was the first probe launched to another planet. But it malfunctioned near Venus while the US Mariner 2 successfully flew past Venus in 1962.

The first spacecraft to reach the surface of another planet was the Soviet space probe called Venera 3. However, it crash-landed on Venus in 1966 while Venera 7 made the first successful landing on Venus in 1970. NASA's Mariner 9 launched in 1971 was the first spacecraft to visit and orbit Mars, and Mars has been significantly explored since then with a multitude of orbiter spacecraft and surface rovers.

The first spacecraft to visit Mercury was called Mariner 10. In 1974-75 it flew by Mercury three times mapping half of its surface and it discovered its thin atmosphere and magnetic field. Launched in 1977, Voyager 2 from the USA passed about 4,950 km above Neptune's north pole in 1989.

NASA's Pioneer 11 was the first spacecraft to provide a close look of Saturn in 1979 and that was followed by Voyager 1 and 2 in 1980 and 1981, respectively. Importantly, Cassini spacecraft started to orbit around Saturn in 2004 and the mission lasted 13 years. Launched in 1989, Galileo was the first spacecraft to orbit Jupiter in 1995 and this Jovian mission lasted 8 years until 2003.

Launched in 2006 the New Horizons spacecraft had a flyby in 2015 of Pluto. Subsequently in 2019, New Horizons also had a flyby of an object in Kuiper Belt, a region beyond the orbit of Neptune. This region is known to contain comets, asteroids and other small bodies, all remnants from 4.5 billion years ago when the solar system was formed and the planets were created.

Of all the planets, Mars has been explored most to date through many orbiting satellites and importantly through four surface rovers. All these robotic vehicles are conceived to transverse the surface of the planet and undertake measurements of its composition, especially to find some evidence of water and life there.

The first rover mission from NASA was called Sojourner and it lasted just a few months in 1997 on Mars. Of the following two rover missions, Spirit operated from 2004 for about 6 years while

Opportunity operated onward from 2004 for almost 15 years. And the fourth rover mission named Curiosity has been operating since 2012. On July 30, 2020 NASA launched the Perseverance rover that will land on Mars in early 2021 to seek signs of life and it will also demonstrate for the first time a helicopter named Ingenuity.

However, being the nearest celestial body to Earth, our Moon has garnered the most attention to date. The first spacecraft to reach moon's surface was the Soviet spacecraft Luna 2 in 1959, followed by Luna 9 in 1966 which became the first spacecraft to soft land through a controlled descent. Subsequently, Luna 10 became the first spacecraft to enter orbit around the moon.

The USA started the Apollo human spaceflight program in 1961 to expressly land humans on our moon's surface and it achieved this goal from 1969 to 1972. Apollo 11 mission succeeded in making Neil Armstrong the first human to walk on the moon on July 20, 1969.

Through the five subsequent Apollo missions, twelve men in all have walked on the moon. Since then, the moon has been observed and mapped, including its poles and the dark side through orbiting satellites from the USA, ESA, China and India. In recent years there has been a renewed international interest to go back to the moon and establish a permanent base there as a launching pad for human visits to Mars.

Spacecraft have visited all planets in the solar system, and even some of the moons of these planets. As such, a good knowledge base has been established of the composition and other characteristics of these celestial bodies. In addition, several flyby missions of comets have been undertaken and, importantly, ESA and Japan have each successfully landed a spacecraft on an asteroid, demonstrating human ingenuity and technological advancement for space explorations.

The most notable recent historical milestone for space voyage, however, is the on-going unbelievably long journey of Voyager 1 and 2. These two spacecraft, launched by the USA in 1977, achieved the unique distinction of having travelled through the entire solar system and continuing to go beyond into interstellar space.

EARTH-ORBITING SPACE LABORATORIES

A number of laboratories in low Earth orbits have been launched over the years, starting with Skylab by the USA in 1973. This satellite operated for 24 weeks at an altitude of 435 km and it had a 3-man crew to perform experiments. Subsequently, the space station MIR was operated by Russia from 1986 to 2001 at an orbit of 358 km.

NASA operated the Shuttle program from 1982 to 2011 that consisted of a fleet of five shuttles named Columbia, Challenger, Discovery, Atlantis and Endeavour. These shuttles undertook a total of 135 manned missions over the years and each of these missions typically lasted a few days starting from shuttle's launch into orbit at an altitude ranging from 304 to 528 km to its glide back to Earth. More recently, China launched and operated two space laboratories called Tiangong 1 and 2 in 2011 and 2016, respectively.

The most notable, operational space laboratory, however, is the International Space Station (ISS), the largest structure assembled in space. It is established and operated by the USA with many partner countries: European Union/ESA, Japan, Canada, and Russia. The first ISS component was launched in 1998 and astronauts started to inhabit the station in 2001.

Ever since then, this Earth-orbiting space facility has been inhabited on a continuous basis and it is projected to be operated at least until 2030. The ISS weighs around 420, 000 kg and with a length of 73 m and width of 109 m it can be viewed from Earth with the naked eye as it makes about sixteen orbits each day at a nominal altitude of 408 km.

The ISS can host up to six astronauts at a time and has the power and other resources to undertake a variety of long duration research activities primarily for life and material sciences under the conditions of microgravity. Part of this research is to understand the long-term impact of microgravity and space environment on humans and thereby to help us prepare for really long duration voyages lasting years that will be required to explore Mars and other celestial bodies in our solar system.

9. SPACE ECONOMY AND BENEFIT

The economy of space and the associated benefits are global in scope, driven in large part that from space looking down there are no political or social boundaries to discern. The earliest impetus for developing space technology and launching assets into space came from the public sector, with a national government trying to showcase a country's prowess.

It is now evident that space also offers a venue to capture the public's imagination and support through exploration of nearby celestial bodies, new scientific discoveries, and importantly innovations for public good from viewing and treating the Earth as a whole.

ECONOMY OF SPACE

Since the first satellite Sputnik in 1958, through more than 5,200 rocket launches, about 7, 500 satellites have been placed in orbits of which around 1,200 are currently in operation. Space economy is ever increasing and is based on the investments from both the public and private sectors for building and operating space assets and the revenues generated though the provision of services furnished by these space-based sensors.

As such, space economy is a global enterprise, encompassing a turn-over of several hundred billion dollars per year. This robust and growing economy has quite a wide span wherein its generators, contributors, users and participants are governments, industry, academia, social institutions, and even general public.

• National Expenditures

According to the 2019 report on Space Economy from the Organization for Economic Co-operation and Development (OECD - a group of 34-member countries), space activities continue to expand globally with an ever-growing public and private sector investments. More than eighty countries now have a space program with the public sector space budgets reaching $75 billion in 2017. Of this total expenditure by governments, the USA budget accounted for more than half, followed by China ($9.3 billion), Japan ($4.2 billion) and France ($2.7 billion).

As such, North America has the largest space budget followed by Europe, Asia and Russia. The USA has the biggest budget by far, split largely among NASA and NOAA, the two civilian agencies. The NASA budget alone is typically $17-20 billion per year allocated for science, operations, and human space explorations while NOAA receives about $2 billion for satellite meteorological systems. Canada in comparison typically spends about $250-400 million per year through its Canadian Space Agency (CSA).

In Europe, the largest portion of expenditures on space activities amounting to some $3-8 billion per year goes to the European Space Agency (ESA). ESA works in collaboration with the European Commission (EC) which has its own space activities focused on applications but it employs ESA to build satellites and ground segments.

The main nations contributing funding to ESA are Germany and France followed by other member states such as Italy, Spain, and the Netherlands. Additionally, each member state of ESA or the European Union (EU) has a national space agency that undertakes its own government-funded space program. For example, of France's $2.7 billion space budget, 36% ($972 million) is for ESA, 3% for EUMETSAT and 61% ($1.64 billion) for its national space program. Whereas 60% ($1 billion) of Germany's $1.7 billion budget goes to

ESA and about 30% to national space activities. Thus, Germany's DLR space agency has a budget of about $600 million per year and it is followed by Italy (ASI-$300 million), the UK ($160 million), and the Netherlands ($145 million).

Russia is historically a leader in space but its budget at about $2.77 billion per year seems to have shrunk since the early days of Sputnik and MIR and now the country appears to concentrate upon launchers such as its commercial launch vehicle, Soyuz. India has a robust space program (ISRO, $1.4 billion per year) as well as South Korea (about $500 million per year), both emerging space nations with a global reach.

• Commerce

Space commerce is an important part of the growing space economy, not just the government expenditures. Space commerce is largely concerned with communications satellites and services, launch vehicles and services, Earth Observations (EO) satellites and data services, devices and applications derived from satellite navigation (GPS), and increasingly space tourism in the future given that private sector investments are driving this consumer-based application.

The first commercial application is based on communications satellites, and it is not only the most mature and established globally, it remains the largest revenue generator as evident by companies and consortiums in Canada, USA, and Europe. Most of the countries have their own communications satellites and most of these are owned and operated by industry. SatCom market is already of several tens of billion dollars per year and it is projected to continue to grow even more.

Another established market is that of commercial launch vehicle services even though practically all of the suppliers are essentially funded directly or indirectly by governments. Launch services are routinely offered by companies in the USA, Europe (France), Russia, China, Japan and India and there are new entrants on the horizon like South Korea and Israel.

With more than 100 launches every year this is a lucrative and growing multi-billion dollars global business. The revenues are dominated by launch of satellites to GEO orbit and this application alone was estimated to be over $1.4 billion in 2017.

Perhaps the biggest growth in space commerce has come about in the sale of EO imagery, with these data sales estimated to be more than $1.6 billion per year to a variety of customers such as defence and civilian agencies as well as industry.

Recognizing a global need for satellite imagery, France (CNES) was first to create a company to sell optical imagery commercially from its SPOT 1 in the early 1980s. Its success resulted in Canada (CSA) creating its own company in 1995 to commercially sell radar images from its RADARSAT 1 satellite.

Images from SPOT and RADARSAT helped to develop a variety of applications and user capabilities thereby creating a commercial global market for space-based images. The LANDSAT series, although established mainly for scientific use, has opened up its image archives to the public and this optical imagery is perhaps most widely used.

There are many other vendors such as from Germany (selling TerraSAR radar imagery), Italy (selling Cosmo-Skymed radar imagery) and India (selling optical and radar imagery) that are also selling data commercially. However, all the vendors of satellite imagery are not just relying upon government furnished satellites and sensors.

Perhaps, the best illustration of the growing and stable commercial market for space-based imagery is that companies have attracted private capital to fund development and launch of their own Earth observations satellites. These suppliers are competing in the market place not just on price but also on providing differentiating and attractive features to customers such as finer spatial resolution, more frequent coverage and real-time data delivery.

Due to the sensitive nature of space technology, government expenditures in each country are largely geared towards building an indigenous space industry. However, there is also a need for companies to not just rely upon government funding but to compete not only nationally but internationally to broaden their customer base.

As in any industry, space business attracts start-ups and innovators and even venture capital but the risks and entrance barriers remain high and also governments continue to be the biggest customers. However, big companies or consortiums have emerged in the USA and Europe, to compete in launch vehicles and design, building and operation of satellites, first to meet governments' needs and subsequently to service commercial market.

Large corporations are in some sense essential for space business given the large capitals that are required, the long duration cycle of design, build and operations, and the inherent risks involved. Space industry while being dominated by big corporations in the USA and Europe is in fact diverse. The small and medium enterprises concentrate on niche or specialized areas and become suppliers to governments and big corporations or provide value-added services to users.

• Education

Given the wide scope and strategic nature of space there are many non-government organizations in the USA and Europe that are active promoters of space and some even ensure that public funds get applied first to science and explorations.

Space is a multi-disciplinary endeavour and university and academia are also important stakeholders as generators of knowledge and technology by attracting and providing a knowledgeable work force. There are some university institutions especially in the USA, Europe and Canada that are well known centres of expertise in space research. Some of these research centres are designing, building and operating their own satellites and space sensors.

There are some academic institutions that cater to education of students through a curriculum focused on space. But by and large space as a subject gets taught as a part of larger established disciplines like electrical, mechanical or aeronautics engineering. There are post graduate degrees in space or aerospace but most of these concentrate on engineering aspects of satellites rather than being multi-disciplinary in scope.

GLOBAL SPACE BENEFITS

There are some who question public expenditures on space and argue this spending can be better applied to solve some pressing problems on Earth such as hunger, poverty and disease. While such an outlook is understandable, public funding for space activities is conducive to generating both tangible (e.g., employment and economic productivity) and intangible (e.g., pride and knowledge) socio-economic benefits to individuals and societies.

It is clear however, that there exists a lack of public exposure to the nature and scope of these benefits and also that these have yet to reach all the world's citizens in an equal and recognizable way.

Space in Daily Use

Space assets are just not for scientific research anymore; these assets are now vital to preserving our quality of life on Earth. Earth-orbiting satellites are part of the essential infrastructure that we rely upon in our daily lives. These satellites are more and more integrated with the terrestrial infrastructure to provide us operational services through a seamless web.

• Weather Watch

The local weather casting of any locale around the world is now based on what our satellites see and record at a given moment. This eye in the sky observes the dynamic weather patterns of any location on Earth in a cohesive and timely manner and provides these measurements as input to weather forecasting models. Without these synoptic, multi-sensor observations, now-casting and any projections ahead into the future whether its hours, days, weeks or months will be deficient and not accurate enough.

An accurate and timely knowledge of weather conditions is crucial to any human activity and as a result of space-based sensors it is possible now to plan ahead, enhance economic productivity and protect life and property from extreme weather conditions.

- Travel

Whether we walk, run, or travel by car, train, boat, ship, and even aircraft, we all rely upon positioning information from geo-positioning satellites. All maps in vehicles are created and updated through the observations and signals from our GPS satellites. These space-based devices help us to go safely from one place to another and they also allow us to save time and fuel. Traffic management is helped through these devices and all autonomous, self-driving vehicles must rely upon them.

- Leisure

News, entertainment (sports and tv shows), telephone, radio, television, all rely upon communication satellites to pass and broadcast audio and video and images and text. Terrestrial communications network is seamlessly merged with space-based equipment to pass audio and video signals from any one location to another location, whether on land, ocean, or in the air. While space-based communications are particularly essential in remote or sparsely populated areas, satellite communications allow the quality and accessibility of reception to be the same practically everywhere in the world. This reliance on satellites is even going to increase in the future as some organizations are implementing plans to provide a space-based web or internet service globally.

• Home Life

Practically any resource we consume and any service we need at home is a beneficiary in some way of space-based assets. Food production and distribution, drugs, remote heath delivery, furniture and textiles, oil and gas, metals and non-metals, material goods that we need, all employ satellite services to some extent.

For example, management of renewable resources such as fisheries, agriculture and forestry and non-renewable resources such as minerals is made more efficient and economic through the use of satellite images. Equally, any distribution of goods is aided by maps and tracking through satellite-based geo-positioning devices. Through communications satellites, it is possible to deliver health care remotely, even surgical procedures are being performed from a distance.

• Work Life

Communications among offices across the globe is enhanced through satellites. It is possible for a geographically distributed team to work around the clock in a timely, routine, and reliable manner. Travelling for business is made more efficient and economic due to weather forecasting and positioning devices. Satellites have shrunk the world business-wise and made it a global village for work life. Space assets are a big contributor to the rapid globalization and inter-dependence of national economies.

• Safety and Security

Management of any disaster around the world through space-based observation sensors and communications services is vital in safeguarding life, property, resources and economy. Each phase of a disaster management cycle from prevention, preparation, on-set or operations, to recovery is aided by satellite images. The space-based support helps to minimize loss or damage during a disaster, to provide rapid recovery after, and is conducive to a robust disaster reduction before a disaster.

Space-based sensors are also vital to ensure that some international treaties or regulations are adhered to and, further, to provide supporting evidence of any violation. As such, satellite surveillance provides the only viable technical means to effectively and reliably monitor any nuclear testing or weapon reduction arrangement, marine or air pollution regulation, offshore international fisheries management, agriculture and forestry production regulation, or Climate-Change target. Moreover, it is to be expected that space assets are also employed by defence agencies around the world to enhance security of their citizens.

Societal Benefits

Space-based services are not only of benefit to an individual, they are also ideal to generate benefits to a society at large whether it is a village, city, region or a nation. From space there are no political or social boundaries and what space has to offer can be accessed equally by all.

There is a growing realization that no single organization or a country has the means to put in place a myriad of space assets useful for diverse applications on Earth. As a result, cooperation in space is essential even though competition appears to be the preferred norm around the world to safeguard sensitive technology and commercial interests.

International cooperation is particularly important in Earth observations given that atmospheric and ocean phenomena cross national borders. It is estimated that more than 50% of climate variables rely on satellite data. Earth observations and weather satellites contribute to monitor these variables as well as support management of land and ocean resources and provide crucial data for weather and climate forecasting models.

There are several international organizations that foster systematic collaboration to achieve common global goals such as the

WMO for weather. One notable organization for Earth Observation is the International Group on Earth Observations (GEO) with the primary mission to achieve societal goals around the world. GEO has particularly identified eight areas where space, especially Earth Observations, is or can be of benefit to global society as a whole.

These societal benefits areas are: Biodiversity and Ecosystem Sustainability, Disaster Resilience, Energy and Mineral Resource Management, Food Security and Sustainable Agriculture, Infrastructure and Transportation Management, Public Health Surveillance, Sustainable Urban Development, and Water Resources Management. GEO fulfills its mission by fostering accessibility to EO data and undertaking showcase projects to demonstrate the value of space derived information in decision making for a sustainable planet.

According to the 2019 OECD Report on Space Economy, returns from investments in space programs are global and multi-faceted. These economic returns range from efficiency gains, cost avoidance and cost savings from space applications like EO and SatCom services to the creation of direct employment.

Space missions provide socio-economic benefits through the efficient use and protection of environment and natural resources, spin-offs from space science research and technology, and the enhancements that result from the myriad of space applications.

Space missions demonstrate the value of STEM (science, technology, engineering, and mathematics) education to young generations and foster multi-disciplinary collaboration. Importantly, they result in the creation of a multitude of direct and indirect employment, practically all are professional jobs. As such, hundreds of thousands of professionals are employed world-wide in the space sector; estimated to be more than 80,000 in the USA alone.

These professionals are engineers, scientists, technicians, teachers, academics, lawyers, economists, financial analysts, and commerce and business majors. Space workers around the globe see themselves as members of a fraternity and together they constitute an important contributor to societal benefits.

10. GLOBAL GOVERNANCE

The governance of space in terms of what gets done in space and how it gets done is a continuing process of discussions among nation states under the umbrella of the United Nations. There are some organizations that foster and promote the scientific utility of space, outside the confines of a national perspective. However, in all the debate and policy formulations at the national and international levels there is generally some awareness of underlying competition among nations and non-pacific use of space.

NATIONAL ORGANIZATIONS

Each country has its own organization structure and regulations to govern its space work. Generally, there is a national space agency to look after civilian space. Military space is typically handled by a nation's defence agency, and there can be some link between the civilian and defence functions or in some cases even a hybrid agency.

Given the wide scope and strategic nature of space, in each country there are also non-government organizations which promote space to public, academic and other knowledge sharing institutions to educate students and undertake research about space, and notably industry to undertake space work and realize commercial benefits.

Understandably, the USA and the European Union (EU) have the most elaborate structures and frameworks to exploit space. The USA has NASA as its national space agency but it also has NOAA that is responsible for operational weather satellites for monitoring Earth's atmosphere and oceans.

NASA is an independent agency while NOAA is part of the US Department of Commerce (DOC) and it is the DOC that is also responsible for commercial exploitation of space such as the sale of high-resolution satellite imagery. Other federal agencies look after frequency allocation, communications satellites and launch vehicle regulations.

Of course, NASA has a long record in space through its early Mercury, Gemini, and Apollo programs to land humans on the moon and bring them back safely to Earth. More recently, NASA has undertaken the space shuttle program and perhaps most importantly is spearheading the on-going International Space Station (ISS)-Freedom.

To manage the ISS, the largest human undertaking in space, NASA has an intergovernmental Memorandum of Understanding (MOU) with all the ISS partners, namely, Europe (ESA/EU), Russia (ROSCOSMOS), Japan (JAXA), and Canada (CSA). The ISS MOU defines the governance framework, respective contributions and sharing of the operational costs.

In addition, NASA has a comprehensive program for the exploration of our moon, planets and their moons, comets, and asteroids. As well, it has research activities for the study of the Cosmos (stars, galaxies, dark matter/energy, and black holes) through space- based instruments. NASA is also very active in space science and Earth science for the study of space environment and for the research on Climate Change.

NASA does its work through its field centres situated across the country and contracting out to industry, academia, and laboratories such as Jet Propulsion Laboratory in California. In the USA, the regulatory aspects for launches and space tourism are handled by Federal Aviation Administration while Federal Communications Commission looks after frequency allocation and associated licences.

Russia has the longest record in space through the old Soviet Union that launched the first satellite, Sputnik in Space in 1958. Now Russia has ROSCOSMOS, a state corporation for its space agency that is a partner in the ISS. It undertakes space exploration and science activities and has a strong astronaut program and launch capability. For many years since the end of the USA's space shuttle program, Russia supplied the Soyuz rocket as a vehicle to ferry astronauts back and forth from the ISS. In 2020, the USA is able again to send astronauts to the ISS from its soil through the launch rocket supplied by a company under contract to NASA.

Europe has ESA as its space agency but it also has the European Commission that looks after European space policy and budgets and executes its own space program. As well, Europe has EUMETSAT for meteorological satellites and observations. Additionally, most countries in Europe have their own national space agencies. For example, Germany has a comprehensive national space program through its space agency-DLR, as does France through CNES, Italy through ASI and the UK through the UK space agency.

Germany is particularly strong in the ISS and EO and France in Launch Vehicles and EO. Europe is autonomous in space, having its own indigenous launch capability, a robust space exploration and science program as well as a comprehensive EO program called Copernicus. As well, Europe works with both the USA and Russia and is an active partner in the ISS through its own astronaut cadre.

China has CNSA and India has ISRO, both space agencies with compressive space programs that rely upon indigenous technology and expertise. China has its own launch vehicles, has sent satellites to the moon, and has initiated its own space station. India has its own launch vehicles as well, has sent satellites to the moon and Mars, and has a strong EO program. South Korea has its own satellites and some launch capability. Argentina and Brazil, each has a space program.

Canada has CSA that undertakes a robust space program emphasizing synthetic aperture radar observations and space science. It is an active participant in the ISS through the contributions of its iconic robotic arm system and a comprehensive astronaut program.

INTERNATIONAL ORGANIZATIONS

Given that space assets are being applied for a variety of purposes from almost all countries in the world there is an increasing challenge as how best to harmonize national initiatives, avoid duplications and put in place a civil regime for operations and use of these assets. These assets are of economic importance that impart both direct and indirect benefits nationally and globally. It is clear as well that while competition is the norm for global order, space is perhaps unique due to the enormity of its scope and potential that it requires both competition and collaboration for a sustainable growth.

There are two international organizations that actively foster sharing of space know-how globally. One is the International Astronautical Federation (IAF) and the other SpaceOps organization. The IAF was founded in 1951 by scientists from ten nations working in space research and it is the oldest organization devoted to sharing of space knowledge.

The IAF has grown into the largest space collective, its membership is comprised of national organizations, academia and private companies from all over the world. It has a broad mission that is geared towards the promotion of cooperation, advancement of international development, raising of awareness and preparing the workforce of tomorrow. To further this mission, it organizes an annual International Astronautical Congress to bring together more than 6,000 participants to share latest space information and developments.

SpaceOps also organizes bi-annual conferences and symposia but its focus is towards technical interchange for space mission operations and ground data systems. It was founded in 1990 and its membership is comprised of thirteen national space agencies plus university organizations and private companies, all working together to promote and maintain an international community of space operations experts.

There are three international organizations particularly dedicated to ensure that Earth Observation satellites get widely utilized globally and that there is a concentrated effort to realize benefits from these space assets for the common good. GEO (Group on Earth Observations) is a partnership of more than 100 national governments and in excess of 100 participating organization, all working together to implement a Global Earth Observation System of System (GEOSS) to better integrate diverse national observing systems.

The mission of GEO is primarily to connect the demand for sound and timely environmental information with the supply of data and information about Earth. As such this intergovernmental group works towards ensuring that benefits of GEOSS are accrued to all members and that the development of an integrated global system is orderly and coordinated.

An operational international organization for utilizing EO is the International Charter Space and Major Disasters. The Charter has fourteen participating members, essentially space agencies that have space-based imaging capability in support of any response to a declared national environmental emergency. The Charter is operational around the clock as it provides priority and rapid access to the satellite assets of its members to observe a particular disaster from its very beginning. These timely satellite images are supplied free of charge to national civil protection agencies whenever there is a natural disaster such as a flood or earthquake anywhere in the world.

In 1979, four nations (Canada, France, the USA, and the Soviet Union) conceived and initiated the International Cospas-Sarsat program to offer satellite-aided search and rescue around the world. Now it is a cooperative of forty-five nations and agencies that is dedicated to detecting and locating radio beacons activated by persons, aircraft or vessels in distress. Utilizing a network of satellites that provide coverage everywhere on Earth, distress alerts are detected and forwarded free of charge to rescue authorities in more than 200 countries.

There are other international organizations, some affiliated with the United Nations (UN) that try to coordinate an orderly development of space. For example, slots for communications satellites in the geostationary orbit—a valuable and competitive resource—are allocated by the International Telecommunications Union (ITU). ITU is a specialized agency of the UN founded in 1865 to facilitate international connectivity in the communications network. As such, ITU also allocates global radio spectrum for a diversity of uses on Earth, including for satellites, so as to avoid or minimize any interference and degradation of service.

United Nations (UN)

Importantly, the UN as a global deliberative body seeks to continually establish an up-to-date international framework for operations in and from space. In 1958, the United Nations General Assembly established the United Nations Office for Outer Space (OOSA) to support its then ad-hoc Committee on the Peaceful Uses of Outer Space (COPOUS).

The UN OOSA fosters international cooperation in space and also implements the responsibilities of the UN under its space law. The international space law is composed of Treaties and Principles and General Assembly Resolutions; five such international treaties plus five sets of principles have been adopted to date.

in 1967, the UN General Assembly adopted the Outer Space Treaty outlining the governance of outer space that includes such provisions as: safeguarding space and its exploration, use and benefits for all mankind, and prohibiting placement of nuclear weapons and militarization of outer space. Then in 1979, the moon Agreement was adopted concerning the resources on the moon, making them a common heritage for all mankind.

In 1974, the Convention on Registration of Objects Launched into Outer Space was adopted by the UN General Assembly. As such, any member state launching in space a rocket, a satellite, or a spacecraft is obliged to register it with the registry maintained by the UN OOSA.

The UN Convention on International Liability for damage caused by space objects was also put in place in 1972 to deal with the repercussions of any debris from space falling on the territory of a member state. The fifth treaty is the Rescue Agreement concerning the rescue and the return of astronauts and the return of objects launched into outer space.

In addition, there are the five UN Principles passed concerning various aspects of space: Law, Broadcasting, Remote Sensing, Nuclear Power Sources, and Benefits. For example, the adopted Remote Sensing principle deals with remote sensing of the Earth from space (i.e., Earth Observations-EO) and it asserts rights of the sensed state. Such sensing information needs to be in the public domain and member states have a right to acquire such imagery of their own territory from a satellite owner and operator on a non-discriminatory basis and on reasonable cost terms.

However, one of the vital emerging topics for space governance is the threat posed by space debris. Avoiding space debris or junk is a growing challenge for any operation in space. This threat is particularly acute in low Earth orbits, i.e., below 1,000 km in altitude, although some space debris is found at all altitudes even in the geostationary orbit, 35,786 km up.

Space debris is the term referring to man-made objects or artificial material orbiting around the Earth that are no longer functional. This space debris is comprised of large objects such as defunct satellites and discarded launch vehicle stages and small pieces like components and particles. Given that these objects are orbiting at a very high speed, up to 8 km per second, means that any collision even with a speck of paint can damage a functioning satellite.

It is estimated that at present in near-Earth space, there are more than 29,000 objects measuring in excess of 10 cm and 167 million objects under 10 cm of which around 200,000 pieces are between 1 and 10 cm.

The objects below an altitude of 600 km continue to orbit for several years before they enter Earth's atmosphere while those above 1,000 km continue to orbit for centuries. All objects eventually pass through Earth's atmosphere and they generally are expected to disintegrate and burnup due to the heat caused by friction.

However, large pieces are known to have survived descent through Earth's atmosphere and have fallen on land or in oceans. Given that the middle of the Pacific Ocean is the most open area on Earth without nearby land, that oceanic region is the preferred location of a graveyard for falling space objects, if their reentry can be controlled.

The amount of space debris is now considered a threat to both manned and unmanned spaceflights. Therefore, operational and routine surveillance of these objects is an increasing priority of space faring nations. Sensors on ground and in space (radars, lidars, optical telescopes) are being applied in a concerted manner to track and identify and determine operational hazards to functioning satellites.

For example, the Combined Space Operations Centre (CSpOC)—the US led multinational space operations centre—manages and operates a Space Surveillance Network. It is reportedly tracking more than 14,000 pieces of space debris larger than 10 cm along with all the functioning satellites in near-Earth orbits.

When there is a potential risk of collision an alert is issued to the impacted space operator so that any effort to change course, if possible, can be undertaken in time. In addition to CSpOC, given the strategic and harmful nature of space debris, Europe, through ESA and national space agencies, is establishing its own space surveillance capability and Russia and China are doing that as well.

Recently, the UN COPOUS has issued guidelines to address the growing challenge of space debris in near-Earth orbits. These guidelines call for the passivation of a satellite post its useful life to avoid its disintegration and to actively bring the satellite down to lower orbits, below 500 km.

The lowering of the altitude is recommended so as to ensure that a satellite enters into Earth's atmosphere as quickly as possible and gets burnt out in less than 25 years. This is a useful practical attempt to limit future growth of space debris. However, there is already planned a quantum increase in the number of small and even micro-satellites in near-Earth orbits and many of these are unlikely to carry fuel to help lower their altitudes.

Moreover, according to one theory, space debris can increase exponentially due to collisions between existing space debris objects themselves, making large pieces breakup into ever smaller pieces. Given the operational risk, there are increasing research efforts to find technological solutions to mitigate the threat, such as a means to remove space debris entirely and to clear-up the congested orbits.

Above all else in space governance, there is the ever-present danger that somehow space will get militarized and become just another theatre for war. For example, there are reports that in recent years some nations have tested anti-satellite weapons and blown-up satellites. These activities have resulted in significantly increasing space debris on one hand and on the other highlighting the fragility of preserving space for peaceful uses.

As always, there also remains the likelihood that the commercial exploitation of space will supersede or weigh over the public good utility of space. The future of space use, especially in terms of what is acceptable commercialization and prohibition of militarization, is increasingly dependent upon the strength of its civic governess structures. Continuing vigilance is required to ensure that the access to space is always open and universal and that benefits from space assets are for the common public good and need to be equally shared around the world.

The role of national and international civil societies is important and relevant as ever to safeguard space for peaceful uses and to see that it is applied for the advancement of all life on Earth. Space ventures help fulfill a common human aspiration for understanding the nature of life on Earth and its place in the Cosmos.

ANNEX A: UNITS, NOMENCLATURE, IMAGES, VIDEOS

UNITS

Speed of Light (C) = 300,000 km per sec (second)
Wave Frequency (f) = number of cycles per sec = Hz (Hertz)
Wavelength = distance travelled in one cycle = f/C
Kilo = 10^3 = 10 raised to power of 3 = 1,000 = thousand
Million = 10^6 = thousand thousand = 1,000,000
Billion = 10^9 = kilo million
G = giga = billion
Trillion = 10^{12} = million million
μ = micro = 10^{-6} = millionth = 1/million
nano = 10^{-9} = billionth = 1/billion
m = metre
cm = centimetre = 10^{-2} = 1/100 = hundredth of metre
mm = millimetre = 10^{-3}= 1/1000 = thousandth of metre
μm = micrometre = 10^{-6} = 1/million = millionth of metre
nm = nanometre = 10^{-9} = 1/billion = billionth of metre
K = Kelvin, the base unit of temperature.
Each unit on this scale is called a Kelvin, rather than a degree.
0 K = -273 degrees Celsius
273 K = 0 degrees Celsius at which water freezes and boils at 100 degree (373 K).

Lightyear = distance light travels in one year = 9.46 trillion km

NOMENCLATURE

USA = United States of America
NASA = National Aeronautics and Space Administration of USA
NOAA = National Oceanic and Atmospheric Administration of USA
CSA = Canadian Space Agency
EU = European Union
EC = European Commission
Eumetsat = European Organization for Meteorological Satellites
CNES = Centre national d'etudes spatiales = Space Agency of France
WMO = World Meteorological Organization
DLR = Space Agency of Germany
ASI = Space Agency of Italy
CNSA = Space Agency of China
Roscosmos = Space Agency of Russia
ISRO = Space Agency of India
UN = United Nations
UNOOSA = United Nations Office for Outer Space Affairs

IMAGES

"The Blue Marble", "Black Marble", "Pale Blue Dot", "Earthrise"
www.nasa.gov/content/blue-marble-image-of-the-earth-from-appollo-17
https://visibleearth.nasa.gov/images/57723/the-blue-marble
www.nasa.gov/multimedia/imagegallery/image_feature_1249.html
https://earthobservatory.nasa.gov/images/79803/night-lights-2012-the-black-marble
https://solarsystem.nasa.gov/resources/536/voyager-1s-pale-blue-dot/

The Big Bang Timeline

https://commons.wikimedia.org/wiki/File:The_History_of_the_Universe.jpg

Planets-International Astronomical Union

www.iau.org/public/images/detail/iau0603a/

Voyager 1& 2 Road Map

www.nasa.gov/feature/goddard/2017/hubble-provides-interstellar-road-map-for-voyagers-galactic-trek/

www.nasa.gov/sites/default/files/thumbnails/image/stsci-h-p1701b-m2000x1087.png

Electromagnetic Radiation Spectrum

https://pages.uoregon.edu/jimbrau/BrauImNew/Chap03/7th/AT_7e_Figure_03_08.jpg

Earth's Atmospheric Layers

www.nasa.gov/mission_pages/Sunearth/science/atmosphere-layers2.html

www.nasa.gov/sites/default/files/images/463940main_atmosphere-layers2_full.jpg

Radar Applications Images-Hardcover Book

RADARSAT-1 Story: A Canadian Satellite by M. E. McGuire - ISBN-10:0992125200

Periodic Table of Chemical Elements

http://www.astronomy.ohio-state.edu/~jaj/nucleo/
https://www.periodni.com/gallery/modern_periodic_table.png

VIDEOS

Our Universe

https://www.sciencealert.com/this-awesome-video-shows-the-scale-of-the-universe-in-the-best-way-possible
https://www.youtube.com/watch?v=GoW8Tf7hTGA
http://www.nationalgeographic.org/video/origins-universe-101/
http://www.nationalgeographic.org/video/space-101-solar-system/
https://www.nytimes.com/interactive/2020/science/exploring-the-solar-system.html?smid=em-share

Pale Blue Dot-Carl Sagan-Humanity Speech

https://youtu.be/EWPFmdAWRZ0
https://www.youtube.com/watch?v=GO5FwsblpT8

Orbital Mechanics

https://letstalkscience.ca/educational-resources/backgrounders/orbital-mechanics
https://youtu.be/BvjlBpP4zU8

ANNEX B: CANADA IN SPACE

Canada was the third country in space after the Soviet Union and USA through the construction of its Alouette 1 satellite that was launched by NASA in 1962. Alouette was designed to study the Ionosphere, to investigate properties of this important atmospheric layer for terrestrial radio communications with respect to their dependence on geographical location, season and time of the day.

Alouette 1 was deliberately switched off after a 10-year mission and the deactivated satellite is still in its circular, 1,000 km altitude orbit. To further study the Ionosphere and the Aurora Borealis-the Northern Lights- Canada also built and launched two satellites under the term International Satellites for Ionospheric Studies, called ISIS I and II in 1969 and 1971, respectively.

Since these early space missions, Canada has had a space program that was fortified and consolidated through the establishment of the Canadian Space Agency (CSA) in 1989. The CSA reports to the Canadian Parliament through the Minister of Innovation, Science and Industry (ISED).

The CSA coordinates Canada Space Program working in collaboration with other federal departments, industry, academia, and provinces. A licence for operating frequency for a satellite operating from Canada is provided by ISED while a licence for operating an Earth Observations (EO) satellite from Canada is handled by Global Affairs Canada.

Prior to the CSA, Canadian interest in space was expressed by a number of different departments with National Research Council responsible for the robotic arm and astronauts, Department of Communications for SatCom, Natural Resources Canada for EO and Environment Canada for meteorological satellites. The Alouette satellite was built and operated by Canadian Deference Research Telecommunications Establishment.

Natural Resources Canada has continued to be strong in EO through its ground stations for receiving EO data at Gatineau, Prince Albert and Inuvik. Environment Canada remains a pioneer user of satellite data and continues to be responsible meteorological satellites.

Canada has focused its programs to address its national needs such as in telecommunications and EO and also be a partner in international ventures especially with the USA-NASA and Europe-ESA.

To look after the communications needs of its population distributed over a vast geographical area Canada was the early adopter of satellite communications. Canada built a Communications Technology Satellite called Hermes and it operated from 1976 to 1979 and became the first communications satellite used for video art and telemedicine such as emergency medical services, teleconferencing and community TV.

In 1969, Canadian government founded Telesat Canada, a Crown Corporation and it launched Anik A1 in 1972 as the world's first domestic communications satellite in geostationary orbit operated by a commercial company. This put Canada in a leadership position in this strategic area given that the company, now private, has launched more than twenty-eight communications satellites over the years and has become one of the largest satellite operators in the world.

To allow surveillance of its long coastline, ice in the Arctic and icebergs along the Labrador Sea and natural resources over its vast land mass, Canada was also an early adopter of Earth Observation from space. In 1995, Canada launched, RADARSAT 1, a satellite that had on board a synthetic aperture radar, an instrument that provides detailed images of the Earth, day/night and in all-weather conditions.

RADARSAT 1 operated until 2012 and it made Canada a global leader in the supply of radar imagery. It was instrumental in creating a commercial market for such satellite observations around the world. The pioneer mission established the RADARSAT brand in more than sixty countries. RADARSAT 2 launched in 2008 is still in operation. And the RADARSAT Constellation Mission (RCM), a constellation of three satellites was launched in 2019 to continue Canadian leadership. Each RCM satellite also carries another sensor called Automatic Identification System (AIS) for ships.

Perhaps the most significant Canadian contribution to EO is SciSat 1, the satellite launched in 2003 to study the concentrations of ozone and other constituents in the Stratosphere over the Arctic using an optical Fourier Transform infrared spectrometer. This satellite designed for a 2-year mission is still in operation and continues to provide such a vital measurement of atmospheric chemicals operationally to the international scientific community.

Other Canadian EO contributions have been through the Optical Spectrograph and InfraRed Imaging System (OSIRIS–a payload on Sweden's Odin Satellite). Launched in 2001, it continues to provide information on the upper atmosphere. Canada also contributed an important payload called Measurements of Pollution in the Troposphere (MOPITT) on board NASA's Terra satellite in 1999. Still operational, its infrared spectroscope measures profiles of carbon monoxide.

Canada launched in 2003 the world's smallest space telescope. It was a micro satellite called the Microvariability and Oscillation of Stars (MOST). Then in 2013 Canada launched the world's first space telescope dedicated to detecting and tracking asteroids and satellites. Still operational, it is called the Near-Earth Object Surveillance Satellite (NEOSSat).

A particularly iconic contribution by Canada in space was the first Canada Arm (called Shuttle Remote Manipulator System) on a USA space shuttle in 1988. Canada has continued its leadership in such robotic manipulators in space through the supply of such arms on all the US space shuttles.

Perhaps more importantly, Canada contributed a more advanced robotic system on the International Space Stations (ISS). It is called the Mobile Servicing System (MSS) and is composed of a remote manipulator known as Canadarm 2 and a special purpose dexterous manipulator, called Dexter but known as Canada hand. Starting in 2001, the MSS helped build the ISS and it keeps the station well maintained and operational in orbit.

Canada has a robust astronaut program, starting with Marc Garneau going into space in 1988 on board NASA's Space Shuttle. Since then nine Canadian astronauts have participated in seventeen missions, on board the Space Shuttles, MIR and ISS, logging more than 15000 hours in space. Three Canadian astronauts have spent extended durations of about six month each on board the ISS with Chris Hadfield also serving as commander of the ISS.

A Canadian instrument has also gone to the surface of Mars. It is called the Alpha Particle X-ray Spectrometer (APXS) designed to perform tests and measure chemical content of Martian soil samples. It has been operating as a part of NASA's Curiosity Mars Rover that was launched in 2011.

Canada has an active space industry with niche expertise not only in design and manufacturing of satellites, payloads, and ground segments but in applications development and value-added services. According to the 2019 report-The Space Economy in Figures by the Organization for Economic Co-operation and Development-OECD-the 2017 budget of the Canadian Space Agency amounted to $249 million.

In 2017, the Canadian space sector generated about $4.1 billion in revenues, 84% of which was from the downstream segment comprised of satellite operations, value-added products and services, and direct-to-home broadcasting. The space sector employed some 9942 full-time equivalents, not counting government employees. Practically all Canadian provinces have some space industry and academic institutions with space expertise.

CPSIA information can be obtained
at www.ICGtesting.com
Printed in the USA
BVHW040145220421
605581BV00006B/62